Moving Broadband Mobile Communications Forward - Intelligent Technologies for 5G and Beyond

Edited by Abdelfatteh Haidine

Published in London, United Kingdom

IntechOpen

Supporting open minds since 2005

Moving Broadband Mobile Communications Forward - Intelligent Technologies for 5G and Beyond
http://dx.doi.org/10.5772/intechopen.83169
Edited by Abdelfatteh Haidine

Contributors
Hao Wu, Valery Tikhvinskiy, Victor Koval, Shashank Kumar Gupta, Jamil Y. Khan, Duy T. Ngo, Lei Chen, Jiajia Chen, Jun Li, Xiaoman Shen, Abdelfatteh Haidine, Salmam Fatima Zahra, Abdelhak Aqqal, Aziz Dahbi, Jupeng Ding, Jiong Zheng, Isiaka Ajewale Alimi, Romil K. Patel, Aziza Zaouga, Nelson J. Muga, Qin Xin, Armando N. Pinto, Paulo P. Monteiro, I. Chih-Lin

First published in London, United Kingdom, 2021 by IntechOpen
IntechOpen is the global imprint of INTECHOPEN LIMITED, registered in England and Wales, registration number: 11086078, 5 Princes Gate Court, London, SW7 2QJ, United Kingdom
Printed in Croatia

British Library Cataloguing-in-Publication Data
A catalogue record for this book is available from the British Library

Additional hard and PDF copies can be obtained from orders@intechopen.com

Moving Broadband Mobile Communications Forward - Intelligent Technologies for 5G and Beyond
Edited by Abdelfatteh Haidine
p. cm.
Print ISBN 978-1-83962-343-1
Online ISBN 978-1-83962-344-8
eBook (PDF) ISBN 978-1-83962-345-5

For EU product safety concerns:
IN TECH d.o.o., Prolaz Marije Krucifikse Kozulić 3, 51000 Rijeka, Croatia, info@intechopen.com or visit our website at intechopen.com.

Meet the editor

Dr.Ing. Abdelfatteh Haidine received his Ph.D. in 2008 from the Technische Universität Dresden, Germany, with a focus on the planning and optimization of telecommunications networks. He worked as a consultant and manager for the deployment of smart metering systems and smart grid applications. Currently, he is an assistant professor for wireless/mobile communications and intelligent systems with the Laboratory of Information Technologies, National School of Applied Sciences, Morocco. His research interests include different issues related to Machine-to-Machine (M2M) and Internet-of-Things (IoT) communications, networking technologies for smart domains: smart maritime port, smart city and smart grid applications, and so on. This covers LPWA networks and their techno-economical aspects. Dr. Haidine also deals with the application of combinatorial optimization as well as the Game Theory paradigm in network planning/migration and resources allocation in broadband mobile networks. In addition, he investigates artificial intelligence and machine learning in optimization procedures/paradigms.

Contents

Preface

The rollout of Long Term Evolution-Advanced (LTE-A) as the fourth generation of mobile communications brought mobile systems to unprecedented throughput of more than 600 Mbps. Therefore, the fifth-generation (5G) of mobile communications had to go beyond the objective of the realization of more transmission capacity. Indeed, 5G targeted the fulfillment of the communications requirements of different vertical application fields. The realized performance covers three-dimensional space, namely, (1) enhanced mobile broadband (eMBB) for bandwidth-hungry applications such augmented reality/virtual reality (AR/VR); (2) ultra-reliable low-latency communication (URLLC) for use cases with high sensitivity to delay like in vehicle-to-vehicle (V2V) or vehicle-to-infrastructure (V2I) applications; and (3) massive Internet-of-Things (IoT). To achieve high performance, 5G relies on several enhanced technologies, which have already shown their potential with LTE-A (4G) and LTE-A Pro (4.5), such as multi-input multi-output (MIMO), spectrum flexibility, higher spectral efficiency, machine-to-machine communication/ machine-type communications (M2M / MTC), and so on. Furthermore, 5G exploits the advantages of softwarization techniques using fog computing, cloud computing, artificial intelligence (AI), machine learning (ML), and more. This book discusses some of these softwarization techniques and some practical aspects from the 5G deployment scenarios. This book is a reference to mobile broadband networks as well as practical use cases of wireless broadband communications. Committed to bridging the gap between theory and practice, this book is also a concise guide for graduate students and readers interested in studying next-generation mobile networks (NGMN) and concepts/applications of mobile communications engineering.

Section 1, "System Realization and Enabling Technologies," presents different computing-based paradigms such as cloud computing and fog computing in the context of 5G. Chapter 1, "Trends in Cloud Computing Paradigms: Fundamental Issues, Recent Advances, and Research Directions toward 6G Fog Networks," presents a comprehensive review of these architectures and their associated concepts. The chapter also discusses beyond 5G (B5G/6G) perspectives. Chapter 2, "Low-Latency Strategies for Service Migration in Fog Computing Enabled Cellular Networks," presents the concepts of fog computing-enabled cellular networks (FeCN), in which computing, storage, and network functions are provisioned closer to the end-users, thus the latency on transport networks can be reduced significantly. In the context of FeCN, the high mobility feature of users brings critical challenges to maintain service continuity with stringent service requirements. Service migration, referred to as transmitting the associated services from the current fog server to the target one, has been regarded as a promising solution to fulfill service continuity during mobility. Chapter 3, "Artificial Intelligence and Machine Learning in 5G and beyond: A Survey and Perspectives," describes the role of AI and ML in 5G and beyond, to build cost-effective and adaptable performing, next-generation mobile networks. The chapter also presents some practical use cases of AI/ML in a network life cycle. The last chapter of this section, "A Brief Overview of CRC Implementation for 5G NR," gives an overview of the cyclic redundancy check (CRC) implementation in 5G New Radio (NR).

It is obvious that the future mobile communications systems will be increasingly complex and heterogeneous, with different networking technologies. Therefore, Section 2, "Practical Aspects from Next Generation Mobile Landscape," presents three use-case scenarios. In chapter 5, "Prospects of 5G Satellite Networks Development," the authors discuss the spectral and technological aspects of 5G satellite networks developments, issues of architecture and role of delays on quality of services of 5G satellite segments, and the possibility of constructing a 5G satellite segment based on distributed and centralized gNB base stations. It also considers issues of satellite payload utilization for bent-pipe and onboard processing technologies in 5G satellite segments. Chapter 6, "An LTE-Direct-Based Communication System for Safety Services in Vehicular Networks," presents a cluster-based LTE sidelink-based V2V multicast/broadcast architecture to satisfy the latency and reliability requirements of V2V safety applications. This architecture combines a new Proximity Service (ProSe) discovery mechanism for sidelink peer discovery and a cluster-based, round-robin scheduling technique to distribute the sidelink radio resources among the cluster members. The last chapter, "Healthcare Application-Oriented Non-Lambertian Optical Wireless Communications for B5G&6G," introduces the healthcare application of optical wireless communication (OWC) based on non-Lambertian beams, and analyzes different application scenarios of OWC based on Unmanned Aerial Vehicles (UAVs), high-speed trains (HST), and unmanned underwater vehicles (UUVs). Finally, it presents research prospects of OWC in B5G and 6G.

To conclude, the editor is grateful to all colleagues who authored the chapters of this book and contributed with valuable references and interesting results related to their current research and applications. We are also grateful to the members of the support team at IntechOpen for their help and professionalism. Special thanks are due to the reviewers for their willingness to review the chapters and provide useful feedback to the authors. We would particularly like to thank our colleague Abdelhak Aqqal for his thorough feedback in order to improve the quality of the publication.

Abdelfatteh Haidine
Laboratory of Information Technologies,
National School of Applied Sciences,
Chouaib Doukkali University,
El Jadida, Morocco

Section 1

System Realization and Enabling Technologies

Chapter 1

Trends in Cloud Computing Paradigms: Fundamental Issues, Recent Advances, and Research Directions toward 6G Fog Networks

Isiaka A. Alimi, Romil K. Patel, Aziza Zaouga, Nelson J. Muga, Qin Xin, Armando N. Pinto and Paulo P. Monteiro

Abstract

There has been significant research interest in various computing-based paradigms such as cloud computing, Internet of Things, fog computing, and edge computing, due to their various associated advantages. In this chapter, we present a comprehensive review of these architectures and their associated concepts. Moreover, we consider different enable technologies that facilitate computing paradigm evolution. In this context, we focus mainly on fog computing considering its related fundamental issues and recent advances. Besides, we present further research directions toward the sixth generation fog computing paradigm.

Keywords: 5G, 6G, cloud computing, edge computing, fog computing, Internet of Things, mobile edge computing

1. Introduction

The need to achieve excellent Quality of Service (QoS) to facilitate effective Quality of Experience (QoE) is one of the notable factors that has brought about substantial evolution in the computing paradigms. For instance, the cloud computing paradigm has been presented to ensure an effective development and delivery of various innovative Internet services [1]. Also, the unprecedented development of various applications and growing smart mobile devices for supporting Internet-of-Things (IoT) have presented significant constraints regarding latency, bandwidth, and connectivity on the centralized-based paradigm of cloud computing [2–4]. To address the limitations, research interests have been shifting toward decentralized paradigms [2].

A good instance of a decentralized paradigm is edge computing. Conceptually, edge computing focuses on rendering several services at the network edge to alleviate the associated limitations of cloud computing. Also, a number of such edge computing implementations such as cloudlet computing (CC), mobile cloud computing (MCC), and mobile edge computing (MEC) have been presented [2, 5–7]. Besides, another edge computing evolution is fog computing. It offers an efficient architecture that mainly focuses on both horizontal and vertical resource distribution in the Cloud-to-Things continuum [2]. In this light, it goes beyond mere cloud extension but serves as a merging platform for both cloud and IoT to facilitate and ensure effective interaction in the system. Nevertheless, these paradigms demand further research efforts due to the required resource management that is demanding and the massive traffic to be supported by the network. For instance, fog nodes are typically equipped with limited computing and storage resources which may prevent them from being a good solution for supporting and meeting requests of large-scale users. Conversely, cloud resources are usually deployed far away from users, which makes cloud servers unable to support services that demand low-latency. Based on this, there is a need for the integration of fog and other cloud-based computing platforms with an effective multiple access technique for efficient resource management across the fog-cloud platform. In this regard, the overall performance can be improved and effective computation offloading can be offered. One of such schemes for performance enhancement is non-orthogonal multiple access (NOMA) [8].

In addition, there have been significant research efforts toward the sixth generation (6G) networks. Also, it is envisaged that various technologies such as device-to-device communications, Big Data, cloud computing, edge caching, edge computing, and IoT will be well-supported by the 6G mobile networks [9]. Meanwhile, 6G is envisioned to be based on major innovative technologies such as super IoT, mobile ultra-broadband, and artificial intelligence (AI) [3, 10]. Besides, it is envisaged that terahertz (THz) communications should be a viable solution for supporting mobile ultra-broadband. Also, super IoT can be achieved with symbiotic radio and satellite-assisted communications. Besides, machine learning (ML) methods are expected to be promising solutions for AI networks [10]. Based on the innovative technologies, beyond 5G network is envisaged to offer a considerable improvement on the 5G by employing AI to automate and optimize the system operation [11].

This chapter presents various evolutions of computing paradigms and highlights their associated features. Also, different related technological implementations are comprehensively discussed. Besides, it presents different models that focus on effective resource allocation across an integrated computing platform for performance enhancement. Moreover, it presents AI as a resourceful technique for the achievement of high-level automation for efficient management and optimization of the 6G fog computing platform. This chapter is organized as follows. Section 2 presents a comprehensive discussion on the evolution of computing paradigms with related concepts and features. Section 3 focuses on the fog architectural model. We discuss the challenges of fog computing and its integration with other computing platforms in Section 4. In Section 5, we present some models for resource allocation in an integrated fog-cloud hierarchical architecture. Section 6 focuses on the trends toward intelligent integrated computing networks and concluding remarks are given in Section 7.

2. Evolution of computing paradigms

This section presents the evolution of computing paradigms. In this regards, related concepts, features, and architectural models are considered.

2.1 Cloud computing

As aforementioned, cloud computing has been in the mainstream of research and has been revolutionizing the information and communication technology (ICT) sector. Based on the National Institute of Standards and Technology (NIST) definition, cloud computing presents an enabling platform that offers ubiquitous and on-demand network access to a shared pool of computing resources such as storages, servers, networks, applications, and services. These interconnected resource pools can be conveniently configured and provisioned with minimal interaction. Besides cost-effectiveness regarding support for pay-per-use policy and expenditure savings, some of the key inducements for the adoption of the cloud computing paradigm are easy and ubiquitous access to applications and data [12].

It is noteworthy that with the cloud computing paradigm, network entities regarding control, computing, and data storage are centralized in the cloud. For instance, storage, computing, and network management functions have been moved to different network places such as backbone IP networks, centralized data centers, and cellular core networks [13]. However, it is challenging for the centralized cloud model to meet the stringent requirements of the emerging IoT. The IoT comprises varieties of computing devices that are connected through the Internet to support a variety of applications and services [2, 13]. In this context, things such as smart meters, tablets, smartphones, robots, wireless routers, sensors, actuators, smart vehicles, and radio-frequency identification (RFID) tags are Internet-connected to ensure a more convenient standard of living [2, 14–16]. Therefore, the centralized-based paradigm offered by cloud computing is insufficient to attend to the stringent requirement of the IoT. Some of the fundamental challenges of the IoT are presented in this section.

2.1.1 Latency requirements

One of the main challenges of the IoT is the associated stringent latency requirements. For instance, a lot of industrial control systems usually require end-to-end latencies of a small number of milliseconds between the control node and the sensor [17, 18]. Examples of such applications are oil and gas systems, manufacturing systems, goods packaging systems, and smart grids. On the other hand, end-to-end latencies below a few tens of milliseconds are required by some time-sensitive (high-reliability and low-latency) IoT applications like drone flight control applications, vehicle-to-roadside communications, gaming applications, virtual reality applications, and vehicle-to-vehicle communications, and other real-time applications. However, these requirements are beyond what a conventional cloud can effectively support [13].

2.1.2 Bandwidth constraints

The unprecedented increase in the number of connected IoT devices results in the generation of huge data traffic. The created traffic can range from tens of megabytes to a gigabyte of data per second. For instance, about one petabyte is been trafficked by Google per month while AT&T's network consumes about 200 petabytes in 2010. Besides, it is estimated that the U.S. smart grid will generate about 1000 petabytes per year. Consequently, for effective support of this traffic, relatively huge network bandwidth is demanded. Moreover, there are some data privacy concerns and regulations that prohibit excessive data transmission. For example, according to ABI Research, about 90% of the generated data by the endpoints should not be processed in the cloud. In this context, it has to be stored and processed locally [13].

2.1.3 Resource-constrained devices

The IoT system comprises billions of objects and devices that have limited resources mainly regarding storage (memory), power, and computing capacity [15]. Based on these limitations, it is challenging for constrained devices to simultaneously execute the entire desired functionality [19]. Besides, it will be impractical to depend exclusively on their relatively limited resources to accomplish their entire computing demands. It will also be cost-prohibitive and unrealistic for the devices to interact directly with the cloud, owing to the associated complex protocols and resource-intensive processing [13]. For example, some constrained medical devices such as insulin pumps and blood glucose meters have to fulfill certain authentication and authorization tasks. Likewise, it has been observed that most of the resource-constrained IoT devices cannot partake in the blockchain consensus mechanisms such as Proof-of-Work (PoW) and Proof-of-Stake (PoS) protocols in which huge processing power is required for the mining process [15].

2.1.4 Intermittent connectivity

It will be challenging for the centralized-based cloud platforms to offer uninterrupted cloud services to systems and devices such as oil rigs, drones, and vehicles with intermittent network connectivity to the cloud resources. As a result, an intermediate layer of devices is required to address the challenges [2, 15].

2.1.5 Information and operational technologies convergence

The advent of Industry 4.0 facilitates the convergence of Information Technology and Operational Technology. In this context, new operational requirements and business priorities are presented. It is noteworthy that safe and incessant operation is of utmost importance in current cyber-physical systems. This is owing to the fact that service disruption can result in a significant loss or dissatisfaction. Consequently, software and hardware update in such sensitive systems is challenging. This calls for a novel architecture that is capable of reducing the system updates [2].

2.1.6 Context awareness

A lot of IoT applications like augmented reality and vehicular networks require access to be able to process local context information such as network conditions and user location. However, owing to the physical distance between central computing and IoT devices, the centralized-based cloud computing implementation is insufficient to support the requirement [2].

2.1.7 Geographical location

The IoT devices are huge in number and are widely distributed over broad geographical areas. These devices require computation and storage services for effectiveness. However, it is challenging to have a cloud infrastructure that can support the entire requirements of the IoT applications [2].

DOI: http://dx.doi.org/10.5772/intechopen.98315

2.1.8 Security and privacy

The present Internet cybersecurity schemes are mainly designed for securing consumer electronics, data centers, and enterprise networks. The solutions target perimeter-based protection provisioning using firewalls, Intrusion Detection

	Computing		**Reference**
	Cloud	**Fog**	
Deployment	Centralized	Distributed[1]	[13, 21–23]
Planning	Demands complicated deployment planning	Demands cautious deployment planning[2]	[13]
Operation	It is controlled and maintained by the expert cloud personnel and operated in designated environments.	The environments are usually determined by customer demands and may require little or no human expert intervention.	[13, 22]
	Usually owned by large companies.	Depending on the size, it could be owned by large or small companies.	
Supported application	Mainly cyber-domain systems.	Cyber-domain and cyber-physical systems.	[13, 22]
	Few seconds round-trip delay-tolerant applications.	Time-critical applications that demand less than tens of milliseconds.	
Connectivity	Work effectively with consistent connectivity.	Can work with intermittent connectivity.	[13]
Latency	High	Low	[21–23]
Storage and computation capabilities	Strong	Weak	[22]
Energy consumption	High	Low	[22]
Bandwidth requirement	High[3]	Low[4]	[13, 22]
Location	Core network	Edge network	[13, 21]
Location awareness	Partially supported	Supported	[22]
Security aspect	Less secure	More secure	[24]
Attack on moving data	High probability	Very low probability	[24]
Client - server distance	Multiple hops	One hop	[25]
Mobility support	Limited	High	[25]

[1]*A distributed or centralized control system can be employed for distributed fog nodes.*
[2]*Some fog deployment is ad-hoc and demands either minimal or no planning.*
[3]*Bandwidth requirement increases with the aggregate volume of generated data by the entire clients.*
[4]*Bandwidth requirement increases with the aggregate volume of filtered data to be sent to the cloud.*

Table 1.
Comparison of main features of cloud and fog computing.

Systems (IDSs), and Intrusion Prevention Systems (IPSs). Besides, based on the associated advantages, certain resource-intensive security functions have been shifted to the cloud. In this regard, they are focusing on perimeter-based protection by requesting authentication and authorization through the clouds. However, the security paradigm is insufficient for IoT-based security challenges.

2.2 Fog computing

To address the centralized-based limitations and for effective support of the IoT devices, edge computing has been presented [2, 14]. Besides, a broader architecture known as fog computing that is based on a distributed scheme has been presented. In the fog paradigm, storage, control, communication, computation, and networking functions are distributed in close proximity to the end-user devices along the cloud-to-things continuum [13].

In addition, to complement the centralized-based cloud platforms in which data, computing functions, and control functions are stored and performed in the cellular core networks and remote data centers, fog stores a significant amount of data and performs considerable functions at or near the end-user. Likewise, instead of routing the entire network traffic over the backbone networks, a considerable amount of networking and communication are performed at or in close proximity to the end-user in fog computing [2, 8, 15, 20]. In this regard, when applications/tasks are offloaded to the neighboring fog nodes rather than a cloud center, fast-response and low-latency services can be offered by fog computing. Besides, the required enormous backhaul burden between the fog nodes and the remote cloud center is alleviated [8, 20].

Cloud and fog are complementing computing schemes. They establish a service continuum between the endpoints and the cloud. In this regard, they offer services that are jointly advantageous and symbiotic to ensure effective and ubiquitous control, communication, computing, and storage, along the established continuum [13]. In **Table 1**, we present the major features of the cloud and the fog to illustrate the advantage of their complements for effective and ubiquitous service delivery along the continuum.

3. Fog architectures and features

Fog computing can enhance the QoS and the efficiency of different use cases. In this context, it can offer noble technical support for cyber-physical system, Mobile Internet, and IoT. This section presents the fog architectural model and the related advantages of fog computing.

3.1 Three-layer architecture of fog

As aforementioned, one of the main features that differentiate fog from cloud computing is that in the former, resources regarding the storage, communication, control, and computation are deployed in proximity to the end-user devices. Moreover, fog architecture can be predominantly centralized, fully distributed, or somewhere amid the two former configurations. Furthermore, fog architecture and supported applications can be implemented in dedicated hardware and software. In addition, fog architecture can also be virtualized to exploit the associated advantages of network virtualization. This will facilitate the execution of the same application wherever it is demanded. In this context, the demand for dedicated applications will be reduced. It can also encourage an open platform in which applications from different vendors can share a common network infrastructure with support for common lifecycle management. Based on this, different applications can be removed, added, updated, deactivated, activated, and configured, to ensure seamless end-to-end services across the continuum [13].

The fog computing architectural model is usually represented by a three-layer architecture that consists of the cloud, fog, and IoT layers [2, 22, 26]. Besides, a broader N-layer reference architecture has been defined by the OpenFog Consortium [2]. This architecture is an improvement on the three-layer architecture. This subsection focuses on three-layer fog architecture and related concepts.

Figure 1 illustrates hierarchical architecture of fog computing with three-layer. This architecture presents a significant extension to cloud computing. In this regard, to bridge the gap between the cloud infrastructure and the end/IoT devices, it offers a transitional layer that is known as *Fog layer*. We expatiate on the associated layers of the architecture in this subsection.

3.1.1 Terminal/IoT layer

The terminal/IoT layer is the layer that is close to the physical environment and end-user. It comprises numerous devices such as mobile phones, tablets, smart vehicles, smartphones, smart cards, drones, sensors, etc. Typically, these IoT devices are usually distributed geographically. Also, their major purpose is to sense feature data of physical events or objects for onward transfer to the upper layer for processing and/or storage. It is noteworthy that certain local processing can also be executed by a number of devices such as smart vehicles, smartphones, and mobile phones that have substantial computational capabilities. After local processing, the resulting data can then be forwarded to the upper layers [2, 22].

3.1.2 Fog layer

The fog layer is normally positioned on the network edge and is the fundamental layer of fog computing hierarchical architecture. The layer comprises a huge number of fog nodes such as fog servers, base stations, switches, routers, access points, and gateways, which are broadly distributed between the cloud and end-user devices [2, 22]. It should be noted that the fog nodes are not only physical network elements but are also logical ones that execute fog computing services [2].

Moreover, the fog nodes can be based on mobile implementation, when deployed on a nomadic carrier, or static, when fixed at a location. With these implementations, end-user devices can suitably connect with appropriate nodes to have access to the required services. Besides, the nodes are connected to the cloud

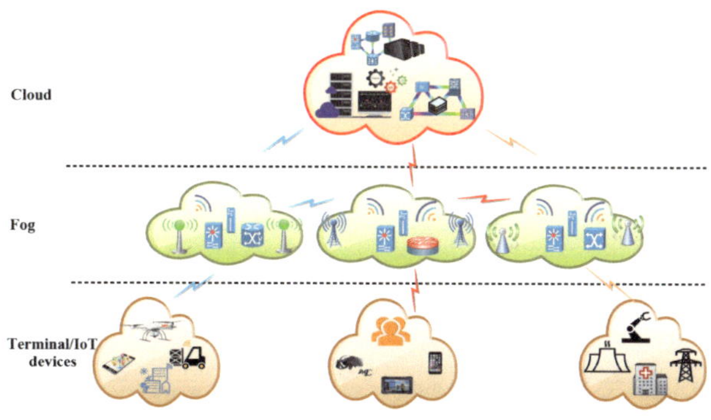

Figure 1.
Fog computing hierarchical network model: a three-tier architecture.

infrastructure through an IP core network for service provision [22]. As aforementioned, fog nodes are capable of computing and transmitting sensed data. Likewise, the received sensed data can be temporarily stored by the nodes. Due to their access to specific network resources; latency-sensitive applications, as well as real-time analysis, can be realized in the fog layer. Furthermore, certain applications demand more powerful storage and computing capabilities. In this context, fog nodes have to interact with the cloud to obtain the required network resources [2, 22].

3.1.3 Cloud layer

The main components of the cloud layer are high-performance servers (with high memory and powerful computational capabilities) and storage devices. Consequently, a huge amount of data can be processed and permanently stored in this layer. Based on this, the layer normally supports various intensive services such as smart factories, smart transportations, and smart homes [2, 22].

Furthermore, unlike the traditional cloud computing architecture, not the entire storage and computing tasks traverse the cloud. Therefore, in the fog architecture, certain services and/or resources can be moved (offloaded) from the cloud to the fog layer through a number of control schemes to optimize resource utilization and increase efficiency [2, 22].

3.2 Features of fog computing

Fog offers various advantages that facilitate new business models and services that can help in expenditure reduction or product rollouts acceleration. In the following subsection, we discuss some of the main advantages of fog network.

3.2.1 Client-centric objective cognizance

Fog architecture consists of widely distributed nodes that support mainly short-range communication and are capable of tracking and getting the end device locations to enable mobility. This feature can facilitate enhanced location-based services and improved potentials for real-time decision making [22]. For example, as fog applications are close to the end-user devices, they can be designed for efficient awareness of the customer requirements. Cognizance of customer requirements helps the fog architecture in establishing the appropriate place to perform storage, computing, and control functions across the cloud-to-thing continuum [13].

3.2.2 Resource pooling and bandwidth efficiency

In fog computing, there can be ubiquitous distributions of resources between the endpoints and the cloud. This helps in the efficient exploitation of the available resources. Besides, with the fog architecture, various applications can leverage the available resources that are abundant but idle on the end-user devices and network edge [13]. For instance, certain computation tasks such as data filtering, data cleaning, data preprocessing, valuable information extraction, redundancy removing, and decision making, are locally performed. In this context, a certain portion of useful data is conveyed to the cloud. Consequently, there is no need to transmit the majority of the data over the Internet. Based on this, fog computing is capable of reducing the network traffic, consequently, the bandwidth is effectively saved [22]. Similarly, the proximity of the fog system to the endpoints facilitate effective integration with the end-user systems. This helps in enhancing the performance and efficiency of the entire system [13].

3.2.3 Scalable and cost-effective architecture

As aforementioned, fog architecture is relatively simple and encourages prompt innovation. Besides, it offers a platform that supports economical scaling. For instance, it is much more cost-effective and faster to use the edge (or client) devices for innovation experimentations instead of that of large operators and vendors networks. In this context, fog encourages an open-market (interoperable) that can support open-application programming interfaces. This is of utmost importance for the proliferation of mobile devices to facilitate innovation, development, deployment, and operation of advanced services [13].

3.2.4 Low-latency and real-time applications

In a fog network, data analytics are allowed at the network edge [13]. The generated data by devices and sensors is acquired locally by the fog nodes at the network edge. The acquired (high priority) data is then processed and stored by edge devices in the local area network. Based on this, traffic across the Internet can be considerably reduced and swift localized services with high-quality can be supported. Hence, time−/latency-sensitive applications for real-time interactions can be supported [22]. For instance, time-sensitive functions can be well-supported for local cyber-physical systems. Besides, this feature is crucial for stability in control systems. Likewise, to support embedded AI applications, the feature is also important for the tactile Internet vision [13]. On the other hand, low priority data that is delay-insensitive can be conveyed to certain aggregation nodes where it will be further processed and analyzed [26].

Furthermore, fog computing focuses on allowing ubiquitous local access to centralized computing resource pools that can be swiftly and flexible provisioned on-demand basis. So, to alleviate communication latency and support delay/jitter sensitive applications, resource-limited end-user devices that are close to the fog nodes can access resource pools. In general, the key native features of fog computing are context awareness and edge location. Besides, it is based on pervasive spatial

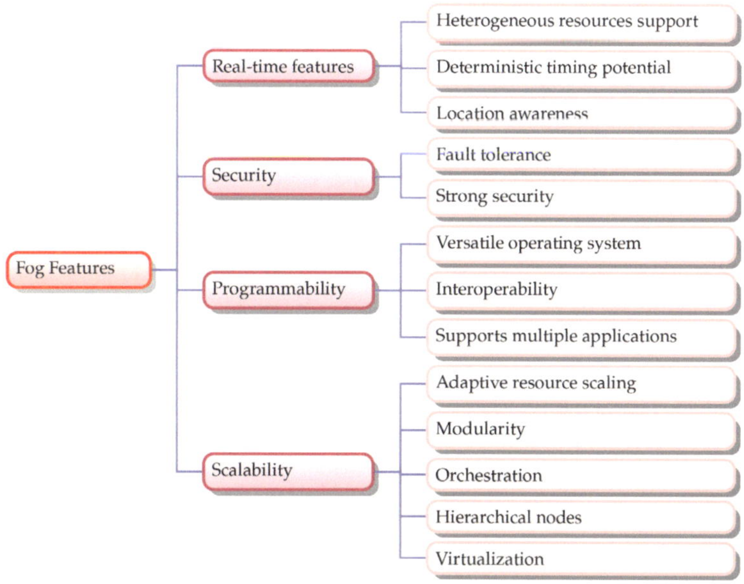

Figure 2.
Fog paradigm features.

deployment to support various devices [27]. In **Figure 2**, we present some of the major features of the Fog paradigm. Also, the following section focuses on the resource allocation challenges in fog computing.

4. Challenges of fog computing and its integration with other computing platforms

As aforementioned, fog computing offers a complementary platform that allows users to offload computational tasks to the network edge. So, resource-limited fog nodes with processing and storage functionalities are deployed at the edge of the network to enhance network performance. Nevertheless, the huge users' data from various applications will be wirelessly transmitted to the cloud center or fog nodes, as the case may be. Based on this, massive communication bandwidth is demanded. It is noteworthy that this is an expensive and constrained resource in the communication systems [8, 28, 29].

Moreover, there are a number of associated challenges of fog-cloud-based computing integrated architecture. One of the main associated challenges is on efficient management of fog infrastructure and allocation of accessible resources to the IoT devices. It is noteworthy that a huge amount of services can be demanded simultaneously by the IoT devices while on the other hand, the respective fog service node is bestowed with limited storage and computing capabilities. Based on this, the entire fog nodes have to be managed optimally. In this context, for the efficient provision of the requested services, they have to be optimally allocated to the required IoT devices. Besides, fog computing resource management is another notable challenge that calls for effective control among fog nodes [26].

Furthermore, it is highly imperative to consider various factors such as energy consumption, service availability, and associated expenses, when fog nodes are deployed to deliver services [30]. In this context, to meet IoT application requirements, optimal mapping of the fog service nodes to the IoT devices is a challenging task. Besides, privacy and security issues such as access control, trust management, intrusion detection, access authentication, etc. [31] in integrated fog computing and IoT device setups are challenging [26].

As aforementioned, one of the main challenges in an integrated computing platform is the multiple resource allocation issue. In this context, efficient resource management across the fog-cloud platform for an effective computation offloading is of paramount importance. To address this challenge, an effective multiple access technique such as NOMA is required [8].

NOMA is an attractive radio multiple access techniques that mainly targets next-generation wireless communications. In NOMA, the power domain for multiple access is leveraged. It presents a number of potential advantages such as high reliability, reduced transmission latency, enhanced spectrum efficiency, and massive connectivity [32]. The main concept of NOMA is to serve multiple users through the same resource regarding time, code/space, and frequency, by exploiting different power levels. Afterward, at the fog nodes, a cancelation technique like successive interference cancelation (SIC) can be implemented to separate and decode the superimposed signals [8, 33]. In the following section, we present some models for resource allocation in an integrated fog-cloud architecture with NOMA implementation.

5. System model

This section presents some related models for the fog computing hierarchical network model illustrated in **Figure 1**. Assume that M fog nodes with various computing and storage capabilities are deployed to offer offloading services given by a set of $\mathcal{M} = \{1, 2, \dots, M\}$. Besides, assume N users denoted by $\mathcal{N} = \{1, 2, \dots, N\}$ with $\mathcal{J} = \{1, 2, \dots, J\}$ independent computation tasks to be executed. The respective task can be expressed as [8]

$$F_{nj} = \left\{A_{in}(nj), Q_{req}(nj), T_{max}(nj), n \in \mathcal{N}, j \in \mathcal{J}\right\}, \tag{1}$$

where $A_{in}(nj)$ is the size of computation input data of the j-th task demanded by the n-th user, $T_{max}(nj)$ represents the maximum tolerable latency of the j-th task required by the n-th user, and $Q_{req}(nj)$ is the total number of central processing unit (CPU) cycles needed to execute the task.

In addition, to express different related models, we assume a quasi-static scenario in which the users are unchanged in the course of computation offloading, however, they can change over different periods. Besides, we assume a perfect instantaneous channel that remains unchanged during the packet transmission. Based on this, we present the following models for an integrated fog-cloud architecture with NOMA implementation.

5.1 Communication model

When an n-th user with a number of offload tasks and the transmission power, p_{mn}, transmits signal, x_{mn}, to the m-th fog node, the received signals, y_{mn}, can be expressed as

$$y_{mn} = \underbrace{\sqrt{p_{mn}} h_{mn} x_{mn}}_{\text{Desired signal}} + \underbrace{\sum_{i \neq n, i \in \mathcal{N}} \sqrt{p_{mi}} h_{mi} x_{mi}}_{\text{Intra-cell interference}} + \underbrace{z_{mn}}_{\text{Noise}}, \tag{2}$$

where the first term represents the desired signal from the n-th user, the second term is the intra-cell interference suffered by the n-th user from other users being served by the m-th fog node on the same frequency band, the third term, z_{mn}, denotes the additive white Gaussian noise (AWGN) with zero mean and variance δ^2, and h_{mn} denotes the channel gain for the n-th user that connects to the m-th fog node.

It is noteworthy that the transmitted signals from various users to each fog node are the desired signals. However, they bring about interference with each other. Also, as individual users that are connected to a specified fog node suffer different channel conditions, the interference can be alleviated and the superimposed signals can be decoded sequentially by each fog node using SIC [8, 33].

In the linear interference cancelation techniques, the desired signal is detected, but other signals are regarded as interference. So, the SIC concept is based on the fact that the signal that has the highest signal-to-interference-plus-noise-ratio (SINR) can be detected first. In this regard, its interference is canceled from other streams [34]. Furthermore, regarding the integrated computing platform, the received signal by a specified fog node from the user that has the highest channel gain is the potential strongest signal, so it is decoded first at the fog node. Afterward, the strongest signal will be removed from the streams. The same approach is then applied to the user with the second-highest channel gain and so on. Consequently, the users' signals on the same frequency band can be sorted in relation to the channel gains. In this context, the users served by the m-th fog node can be arranged in descending as [8]

$$|h_{m1}|^2 \geq |h_{m2}|^2 \geq \cdots \geq |h_{mN}|^2 \ \forall n \in \mathcal{N} \tag{3}$$

Using Eq. (3), every single fog node can subtract and decode the desired signals. Besides, the received SINR, γ, of the n-th user being served through the m-th fog node can be defined as [8, 35, 36]

$$\gamma_{mn}(p_{mn}) = \frac{p_{mn}|h_{mn}|^2}{\delta^2 + \sum_{i=n+1}^{N} p_{mi}|h_{mi}|^2} \tag{4}$$

Furthermore, the resultant transmitted data rate of the n-th user at the m-th fog node can be expressed as [8, 35, 37]

$$R_{mn}(w_{mn}, p_{mn}) = w_{mn} \log_2(1 + \gamma_{mn}(p_{mn})), \tag{5}$$

where w_{mn} represents the occupied frequency band of the n-th user that is served by the m-th fog node and $\mathcal{W}$ denotes the total frequency band.

Moreover, as a result of the limited resources in the fog node, it is challenging to concurrently fulfill the entire services demanded by the end users. So, to acquire the demanded services, each end-user should have a satisfaction function for the evaluation of the allocated resources, ξ. The associated satisfaction function, χ, can be defined as [26]

$$\chi(\xi) = \begin{cases} \log(\xi + 1), & 0 \leqslant \xi < \xi_{\min} \\ \log(\xi_{\max} + 1), & \xi \geqslant \xi_{\min} \end{cases}, \tag{6}$$

where $\xi_{\max}$ denotes the maximum resource that is required to offer the demanded service.

Moreover, based on the satisfaction function, the major objective of the fog node is to offer a global satisfaction maximization for the entire end users. This can be expressed as [26]

$$\textbf{Objective}.\ \max\{\chi_g\} \tag{7}$$

$$\textbf{S.t.}$$

$$\begin{cases} \chi_g = \sum_{i=1}^{n}\{\tau_i \cdot \chi_i(\xi_i)\} \\ \xi_1 + \xi_2 + \cdots + \xi_n \leqslant \Xi \\ \tau_1 + \tau_2 + \cdots + \tau_n = 1 \\ \xi_1, \xi_2, \ldots, \xi_n \geqslant 0 \end{cases}, \tag{8}$$

where χ_g denotes the overall satisfaction of the entire end users, ξ_i denotes the allocated resource to the n-th end-user, Ξ represents the possessed resource by the fog node, and τ represents the associated priority level for the n-th end-user.

Furthermore, using Eqs. (7) and (8), resources of the fog node can be allocated to the entire end-device while the overall maximum satisfaction is achieved. Moreover, the fog nodes are connected and are capable of sharing their resources to deliver the requested service by the end-users. Assume a scenario in which a fog node does not possess sufficient resources to offer services that are locally requested, then it can shift certain requested services with low priority level to the neighboring fog nodes with spare resources for processing. The spare resources, R_s^f, of the m-th fog node can be defined as

$$\Xi_{\text{spare}}^f = \Xi^f - \sum_{i=1}^{n} \xi_i^{\max}, \tag{9}$$

where, $\xi_i^{\max}$ denotes the maximum resource required by the n-th end-user and Ξ^f represents the resource m-th fog node.

5.2 Fog computing model

The fog computing model is based on the required tasks and the associated overhead. For instance, each fog node will receive task offloading requests from the users. Based on its resource capabilities, the respective node is expected to process the requested computational tasks. As a result of this, certain overhead regarding time and energy will be incurred to transmit and process at the fog nodes [8, 37]. The associated overheads are discussed in the following subsections.

5.2.1 Task processing latency

Based on the communication model presented in subsection 5.1, the transmission latency regarding computation offloading can be evaluated. Assume that the m-th fog node receives computation task F_{nj} from the n-th user, the incurred transmission latency when the n-th user send data to offload the j-th task using Eq. (5) can be expressed as

$$T^f_{mnj,t} = \frac{A_{in}(nj)}{R_{mn}} \tag{10}$$

As aforementioned, each fog node possesses limited computation capabilities. Assume that the m-th fog node with computing capability, C^f_{mn}, is assigned to the n-th user, the related computation execution time $T^f_{mnj,e}$ can be defined as

$$T^f_{mnj,e} = \frac{Q_{req}(nj)}{C^f_{mn}} \tag{11}$$

Furthermore, consider a scenario in which each fog node is equipped with a CPU that is based on non-preemptive allocation. Also, assume that computing resource is assigned to an individual user each time until its required tasks are accomplished. Moreover, assume the process sequence, $\mathbf{q}_m = \{q_{ms}|q_{ms} \in \{1, 2, \dots, N\}, q_{ms} \neq q_{mn}, s, n \in \mathcal{N}\}$ in the m-th fog node in which the tasks are executed in the ascending order, q_m. In this scenario, for task j, the queuing delay time can be defined as

$$T^f_{mnj,q} = \sum_{s, q_{ms} < q_{mn}}^{N} b_{ms} T^f_{msj,e} \tag{12}$$

where b_{ms} represents the outcome of user scheduling for the fog nodes that specifies selection of the m-th fog node by the s-th user for offloading. The selection criteria of the m-th fog node by the s-th user can be defined as [8, 37]

$$b_{ms} = \begin{cases} 1 & \text{if } s\text{-th user selects (associated with) the } m\text{-th fog node} \\ 0 & \text{otherwise} \end{cases} \tag{13}$$

The aggregate latency incurred by the fog computing when the n-th user offloads the j-th task to the m-th fog node can be defined as [8, 36, 37]

$$T^f_{mnj} = T^f_{mnj,t} + T^f_{mnj,e} + T^f_{mnj,q} \tag{14}$$

5.2.2 Energy consumption

The energy consumptions for transmitting and processing tasks are the main offloading energy consumptions [30]. When an n-th user offloads j-th task to an m-th fog node, the associated energy consumption can be defined as

$$E^f_{mnj} = \underbrace{\overbrace{E^f_{mnj,t}}^{\text{Transmission energy consumption}} + \overbrace{E^f_{mnj,e}}^{\text{Computing energy consumption}}}_{\text{Off loading energy consumption}} \tag{15}$$

$$= T^f_{mnj,t} p_{mn} + \eta_m C^f_{mn} T^f_{mnj,e},$$

where η_m represents the coefficient that signifies the energy consumption per CPU cycle of an m-th fog node. The first and the second terms are the transmission and computing energy consumptions on the m-th fog node.

5.3 Cloud computing model

As aforementioned, the fog nodes have relatively limited resources regarding memory (storage), power, and computing capacity [15]. Therefore, when resource-limited fog nodes are not capable of accomplishing the requested computational tasks due to their constrained resources, the tasks have to be sent to the cloud center via the backhaul links [8].

Furthermore, the tasks can be efficiently executed at the cloud centers owing to their sufficiently high resource capabilities. It should be noted that additional overhead regarding energy and time will be incurred in the process of forwarding tasks to the remote cloud center [8]. The related overheads are considered in the following subsections.

5.3.1 Task processing latency

Suppose the backhaul links data rate, R^b_m, is available between the m-th fog node and the remote cloud center with computation capability given by C_c. Besides, based on sufficient resources and powerful computation capabilities of the cloud center, the tasks from various users can be executed instantly. In this context, the queuing delay can be omitted in the processing latency analysis of cloud computing. Following the fog computing model analysis presented in subsection 5.2, the aggregate latency that will be incurred in forwarding the j-th task of the n-th user from the m-th fog node to the remote cloud center can be expressed as [8, 35]

$$\begin{aligned} T^c_{mnj} &= T^f_{mnj,t} + T^c_{mnj,t} + T^c_{mnj,e} \\ &= \frac{A_{in}(nj)}{R_{mn}} + \frac{A_{in}(nj)}{R^b_m} + \frac{Q_{req}(nj)}{C_c} \end{aligned} \tag{16}$$

5.3.2 Energy consumption

The aggregate energy consumption of an m-th fog node that offloads a j-th task of an n-th user to a remote cloud center can be defined as [8, 36]

$$\begin{aligned} E^c_{mnj} &= E^f_{mnj,t} + E^c_{mnj,t} + E^c_{mnj,e} \\ &= T^f_{mnj,t} p_{mn} + T^c_{mnj,t} p^b_m + \eta_c T^c_{mnj,e} C_c \end{aligned} \tag{17}$$

where η_c represents a coefficient that signifies the energy consumed by the CPU of the cloud per cycle, p^b_m denotes the allocated power for tasks forwarding to the cloud center by an m-th fog node.

6. Trends toward intelligent-based fog computing

This section focuses on the current trends toward intelligent integrated computing networks. As aforementioned, different challenges as regards scalability, management, and optimization have been presented in the fog-cloud-based computing integrated architecture. For efficient management of resource-limited fog nodes and optimization of the cloud computing platform, the trend is toward the adoption of AI-enabled techniques in the *Network 2030* (6G and Beyond) [4, 10]. For instance, apart from intelligent driving that the 6G network is anticipated to support, it will also offer a promising path toward the industrial revolutions where the future intelligent factories are anticipated to support densely concentrated intelligent mobile robots. Based on this, a number of new service classes like ultra-high data density (uHDD), ubiquitous mobile ultra-broadband (uMUB), and ultrahigh-speed-with-low-latency communications (uHSLLC) have been defined [38, 39]. A typical instance of an integrated hierarchical computing platform supported by the AI techniques is illustrated in **Figure 3**.

Furthermore, in the considered integrated hierarchical computing AI-enabled platform, we consider the fog radio access networks (F-RANs) as a case study. In this context, we discuss how F-RANs can facilitate the deployment of hierarchical AI in wireless networks. Besides, consideration is given to the influences of AI in making F-RANs smarter in rendering better services to mobile devices.

In addition, regarding the influences of F-RANs on the deployment of AI (F-RAN-Enabled AI), the F-RANs present hierarchical layers (cloud, fog, and IoT) that can be exploited. So, F-RAN offers heterogeneous processing capability that can be leveraged for hierarchical intelligence across the integrated layers through centralized, distributed, and federated learning. Besides, to significantly alleviate the memory issue of mobile devices, cross-layer learning can also be employed. Besides, concerning the influences of AI on the F-RANs (AI-Enabled F-RAN), AI presents F-RANs with techniques and technologies for effective support of the huge traffic. Likewise, it helps in making intelligent decisions in the networks. These features can be harnessed through the implementation of ML tools such as reinforcement learning (RL) algorithms and deep neural networks (DNNs) [28, 29, 40, 41]. For instance, DNNs can be adopted for data processing. Besides, RL algorithms can be employed for optimizations and decisions [40]. We expatiate on the relationship between the F-RANs and AI in the following subsections.

6.1 F-RAN-enabled AI

The F-RAN heterogeneous platforms with varied memory and computational resources offer hierarchical application scenarios for the AI. Based on this, hierarchical intelligence such as cloud intelligence, fog intelligence, and on-device intelligence, can be achieved across the layers [40]. In this part, we present learning-based intelligence schemes for the F-RAN-Enabled AI.

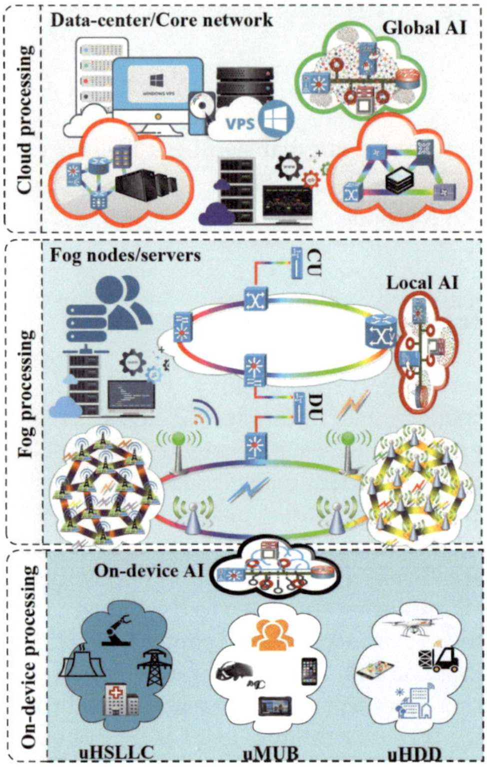

Figure 3.
A typical AI/fog-enabled computing architecture. uHSLLC: Ultrahigh-speed-with-low-latency communications; uMUB: ubiquitous mobile ultra-broadband; and uHDD: ultra-high data density.

6.1.1 Centralized learning-based cloud intelligence

The centralized cloud is not only endowed with considerable access to a global dataset but also has a significant amount of storage and computing power. Consequently, with sufficient data samples, training of the centralized DNN algorithms can be used to leverage the powerful cloud intelligence. Owing to its flexibility and pay-as-you-go capability, cloud intelligence-based services can be on-demand. In this regard, it can scale in accordance with the subscribers' requirements [42].

Moreover, in centralized-based learning, it is assumed that the mobile devices transmit data to the central cloud. However, this is at the expense of communication overhead, regarding bandwidth and energy. Usually, it is challenging to meet the real-time application demands because of the incurred latency. Besides, due to privacy concerns of the mobile devices, attention should also be paid to the transmission of the generated data. Therefore, these concerns demand alternative solutions. A viable approach is based on the exploitation of the distributed architecture and processing capabilities of the mobile devices and/or fog nodes in the development of distributed ML techniques [40, 42].t

6.1.2 Distributed learning-based fog intelligence

In edge computing, cloud resources are leveraged by the fog nodes. Also, the service latency is significantly reduced because the fog nodes are in proximity to the devices. Also, the proximity helps in enhancing privacy. Moreover, based on the distributed natures of the fog nodes and mobile devices, the edge ML algorithms are normally implemented distributedly. In this regard, the training samples are distributed randomly over a considerable number of mobile devices and fog nodes. Each fog node executes training tasks using the gathered local data samples from the associated mobile devices. In addition, when the local model state information (MSI) is aggregated, a global model can be acquired from the fog nodes. Nevertheless, devices in this scheme (distributed ML) have to send their respective data to the fog nodes. This procedure can also violate data privacy. In view of this, federated learning can be employed to further enhance privacy [11, 40, 43].

6.1.3 Federated learning-based on-device intelligence

There have been unprecedented improvements in the processing capability of mobile devices, making joint training and inference more viable. The learning in this layer can be achieved using federated learning architecture [44]. This approach is contingent on periodic computation and exchange of updated MSI versions of the individual mobile devices. Consequently, rather than sending the raw data, the MSI computed using their datasets are exchanged. At the fog node, the distributed MSI updates of the associated mobile devices are aggregated. Based on this, the results of low-latency inference can be acquired. These results can be used in delay-sensitive applications to make a fast response to local events. Apart from being a swifter solution, this approach also helps in enhancing data privacy and is a promising solution for privacy-sensitive applications. In addition, it is also possible to aggregate the distributed MSI updates of the fog nodes to achieve a global model in the cloud. It is noteworthy that this learning process is appropriate for training a low-weight AI model in which there are fewer parameters on mobile devices. Nevertheless, for a remarkably large number of mobile devices and an AI model with huge parameters, wireless data aggregation of MSI updates is challenging [11, 40, 43].

6.1.4 Cross-layer learning-based hierarchal intelligence

In a scenario where the AI model size is more than the memory size of the mobile devices, the mobile devices will be unable to complete the entire model training on themselves. In such a case, the model should be partitioned into sections and distributed over the network entities. So, the lower layers' (mobile devices) outputs will be aggregated prior to transmission to the fog nodes (intermediate layers). Likewise, the intermediate layers' output will be aggregated ahead of transmission to the top cloud layers. One of the advantages of cross-layer learning is that it aids system scalability. Besides, the demand for mobile devices' memory size can also be reduced. Nevertheless, cross-layer DNNs demand stringent training algorithms [40].

6.2 AI-enabled F-RAN

The significant growth in the bandwidth demands by the radio access networks is mainly due to the proliferation of mobile devices and various supported bandwidth-intensive multimedia applications. This results in a traffic explosion that is challenging for the current mobile networks. To address the issue, various network architectures that exploit different types of resources have been presented. However, resource management in such emerging architectures is very demanding. Based on this, innovative techniques for excellent data processing and efficient network optimization are required [20, 40]. The following subsections present AI techniques as viable tools for attending to the associated network challenges.

6.2.1 Intelligent data processing

Based on the growing increase in the diversity of F-RAN applications, the envisaged multimedia data to be supported will be heterogeneous, huge, and high-dimensional. Therefore, direct raw data transmission to the cloud and fog node will bring about high communication overhead. Besides, direct utilization of raw data for network optimization can cause high-computing overhead and low-efficiency issues. Moreover, there has been considerable advancement in the DNNs that facilitate data processing. For instance, convolutional operations have been exploited by convolutional neural networks (CNNs) for spatial feature extraction from input signals [40].

6.2.2 Intelligent network optimization

There are a number of ML techniques such as unsupervised, supervised, and RL algorithms that can be employed for efficient network optimization. For instance, as supervised learning focuses on mapping inputs to outputs in accordance with the training samples, the DNN-based supervised learning is an attractive scheme for beamforming design and power control of fog nodes. On the other hand, unsupervised learning is based on inferring the underlying data structure without any label, so it is appropriate for empirical analysis such as computation offloading, clustering, and resource allocation, in the F-RAN. Besides, in the RL, to maximize predicted cumulative return, sequential actions are taken by actor/agent based on the environment observations [42, 45].

7. Conclusion

In this chapter, we have presented a comprehensive overview of the evolution of computing paradigms and have highlighted their associated features. Moreover, different models that focus on effective resource allocation across an integrated computing platform have been presented. Besides, a comprehensive discussion on efficient resource management and optimization of the 6G fog computing platform to meet strict on-device constraints, reliability, end-to-end latency, bit-rate, and security requirements have been presented. In this context, we have presented AI as a resourceful technique for the achievement of high-level automation in the integrated computing heterogeneous platform.

Acknowledgements

This work is supported by the European Regional Development Fund (FEDER), and Internationalization Operational Programme (COMPETE 2020) of the Portugal 2020 (P2020) framework, under the projects DSPMetroNet (POCI-01-0145-FEDER-029405) and UIDB/50008/2020-UIDP/50008/2020 (DigCORE). It is also supported by the Project 5G (POCI-01-0247-FEDER-024539), SOCA (CENTRO-01-0145-FEDER-000010), ORCIP (CENTRO-01-0145-FEDER-022141), and RETIOT (POCI-01-0145-FEDER-016432).

Author details

Isiaka A. Alimi[1*], Romil K. Patel[1,2], Aziza Zaouga[1], Nelson J. Muga[1], Qin Xin[3], Armando N. Pinto[1,2] and Paulo P. Monteiro[1,2]

1 Instituto de Telecomunicações and University of Aveiro, Portugal

2 Department of Electronics, Telecommunications and Informatics, University of Aveiro, Portugal

3 Department of Science and Technology, University of the Faroe Islands, Tórshavn, Faroe Islands

*Address all correspondence to: iaalimi@ua.pt

References

[1] Michael Armbrust, Armando Fox, Rean Griffith, Anthony D. Joseph, Randy Katz, Andy Konwinski, Gunho Lee, David Patterson, Ariel Rabkin, Ion Stoica, and Matei Zaharia. A View of Cloud Computing. *Commun. ACM*, 53 (4):50–58, April 2010.

[2] M. De Donno, K. Tange, and N. Dragoni. Foundations and Evolution of Modern Computing Paradigms: Cloud, IoT, Edge, and Fog. *IEEE Access*, 7: 150936–150948, 2019.

[3] P. M. Shakeel, S. Baskar, H. Fouad, Gunasekaran Manogaran, V. Saravanan, and Q. Xin. Creating Collision-Free Communication in IoT with 6G Using Multiple Machine Access Learning Collision Avoidance Protocol. *Mobile Networks and Applications*, pages 1–12, 2020.

[4] A. Marahatta, Q. Xin, *C. chi*, F. Zhang, and Z. Liu. PEFS: AI-driven Prediction based Energy-aware Fault-tolerant Scheduling Scheme for Cloud Data Center. *IEEE Transactions on Sustainable Computing*, pages 1–1, 2020.

[5] Y. Liu, J. E. Fieldsend, and G. Min. A Framework of Fog Computing: Architecture, Challenges, and Optimization. *IEEE Access*, 5:25445–25454, 2017.

[6] K. Dolui and S. K. Datta. Comparison of edge computing implementations: Fog computing, cloudlet and mobile edge computing. In *2017 Global Internet of Things Summit (GIoTS)*, pages 1–6, 2017.

[7] Isiaka Ajewale Alimi, Nelson Jesus Muga, Abdelgader M. Abdalla, Cátia Pinho, Jonathan Rodriguez, Paulo Pereira Monteiro, and Antonio Luís Teixeira. *Towards a Converged Optical-Wireless Fronthaul/Backhaul Solution for 5G Networks and Beyond*, chapter 1, pages 1–29. John Wiley & Sons, Ltd, 2019.

[8] Y. Liu, F. R. Yu, X. Li, H. Ji, and V. C. M. Leung. Distributed Resource Allocation and Computation Offloading in Fog and Cloud Networks With Non-Orthogonal Multiple Access. *IEEE Transactions on Vehicular Technology*, 67 (12):12137–12151, 2018.

[9] C. L. Stergiou, K. E. Psannis, and B. B. Gupta. IoT-based Big Data secure management in the Fog over a 6G Wireless Network. *IEEE Internet of Things Journal*, pages 1–1, 2020.

[10] L. Zhang, Y. Liang, and D. Niyato. 6G Visions: Mobile ultra-broadband, super internet-of-things, and artificial intelligence. *China Communications*, 16 (8):1–14, 2019.

[11] I. Tomkos, D. Klonidis, E. Pikasis, and S. Theodoridis. Toward the 6G Network Era: Opportunities and Challenges. *IT Professional*, 22(1):34–38, Jan 2020.

[12] S. U. Khan. Elements of Cloud Adoption. *IEEE Cloud Computing*, 1(1): 71–73, 2014.

[13] M. Chiang and T. Zhang. Fog and IoT: An Overview of Research Opportunities. *IEEE Internet of Things Journal*, 3(6):854–864, 2016.

[14] I. A. Alimi, A. Tavares, C. Pinho, A. M. Abdalla, P. P. Monteiro, and A. L. Teixeira. *Enabling Optical Wired and Wireless Technologies for 5G and Beyond Networks*, chapter 8, pages 177–199. IntechOpen, London, 2019.

[15] Hany F. Atlam and Gary B. Wills. Chapter Three - Intersections between IoT and distributed ledger. In Shiho Kim, Ganesh Chandra Deka, and Peng Zhang, editors, *Role of Blockchain Technology in IoT Applications*, volume 115 of *Advances in Computers*, pages 73 – 113. Elsevier, 2019.

[16] I. Alimi and A. Shahpari and A. Sousa and R. Ferreira and P. Monteiro and A. Teixeira. *Challenges and Opportunities of Optical Wireless Communication Technologies*, chapter 2. IntechOpen, London, 2017.

[17] Z. Ma, M. Xiao, Y. Xiao, Z. Pang, H. V. Poor, and B. Vucetic. High-Reliability and Low-Latency Wireless Communication for Internet of Things: Challenges, Fundamentals, and Enabling Technologies. *IEEE Internet of Things Journal*, 6(5):7946–7970, 2019.

[18] M. Weiner, M. Jorgovanovic, A. Sahai, and B. Nikolié. Design of a low-latency, high-reliability wireless communication system for control applications. In *2014 IEEE International Conference on Communications (ICC)*, pages 3829–3835, 2014.

[19] Corinna Schmitt, Claudio Anliker, and Burkhard Stiller. Chapter 8 - Efficient and Secure Pull Requests for Emergency Cases Using a Mobile Access Framework. In Quan Z. Sheng, Yongrui Qin, Lina Yao, and Boualem Benatallah, editors, *Managing the Web of Things*, pages 229 – 247. Morgan Kaufmann, Boston, 2017.

[20] I. A. Alimi, A. L. Teixeira, and P. P. Monteiro. Toward an Efficient C-RAN Optical Fronthaul for the Future Networks: A Tutorial on Technologies, Requirements, Challenges, and Solutions. *IEEE Communications Surveys Tutorials*, 20(1):708–769, 2018.

[21] Z. Liu, J. Zhang, Y. Li, L. Bai, and Y. Ji. Joint jobs scheduling and lightpath provisioning in fog computing micro datacenter networks. *IEEE/OSA Journal of Optical Communications and Networking*, 10(7):152–163, 2018.

[22] Pengfei Hu, Sahraoui Dhelim, Huansheng Ning, and Tie Qiu. Survey on fog computing: architecture, key technologies, applications and open issues. *Journal of Network and Computer Applications*, 98:27 – 42, 2017.

[23] E. Baccarelli, P. G. V. Naranjo, M. Scarpiniti, M. Shojafar, and J. H. Abawajy. Fog of Everything: Energy-Efficient Networked Computing Architectures, Research Challenges, and a Case Study. *IEEE Access*, 5:9882–9910, 2017.

[24] C. Nandyala and Haeng-Kon Kim. From Cloud to Fog and IoT-Based Real-Time U-Healthcare Monitoring for Smart Homes and Hospitals. *International Journal of Smart Home*, 10: 187–196, 2016.

[25] Syed Noorulhassan Shirazi, Antonios Gouglidis, Arsham Farshad, and David Hutchison. The Extended Cloud: Review and Analysis of Mobile Edge Computing and Fog From a Security and Resilience Perspective. *IEEE Journal on Selected Areas in Communications*, 35(11):2586–2595, 2017.

[26] J. Lin, W. Yu, N. Zhang, X. Yang, H. Zhang, and W. Zhao. A Survey on Internet of Things: Architecture, Enabling Technologies, Security and Privacy, and Applications. *IEEE Internet of Things Journal*, 4(5):1125–1142, 2017.

[27] E. Baccarelli, M. Scarpiniti, and A. Momenzadeh. Fog-Supported Delay-Constrained Energy-Saving Live Migration of VMs Over MultiPath TCP/IP 5G Connections. *IEEE Access*, 6: 42327–42354, 2018.

[28] Isiaka A. Alimi, Paulo P. Monteiro, and António L. Teixeira. Analysis of multiuser mixed RF/FSO relay networks for performance improvements in Cloud Computing-Based Radio Access Networks (CC-RANs). *Optics Communications*, 402:653 – 661, 2017.

[29] I. Alimi, P. Monteiro, and A. Teixeira. Outage probability of multiuser mixed rf/fso relay schemes for heterogeneous cloud radio access

networks (h-crans). *Wireless Personal Communications*, 95:27–41, 2017.

[30] Isiaka Ajewale Alimi, Abdelgader M. Abdalla, Akeem Olapade Mufutau, Fernando Pereira Guiomar, Ifiok Otung, Jonathan Rodriguez, Paulo Pereira Monteiro, and Antonio Luís Teixeira. *Energy Efficiency in the Cloud Radio Access Network (C-RAN) for 5G Mobile Networks*, chapter 11, pages 225–248. John Wiley & Sons, Ltd, 2019.

[31] Zhijiang Chen, Guobin Xu, Vivek Mahalingam, Linqiang Ge, James Nguyen, Wei Yu, and Chao Lu. A cloud computing based network monitoring and threat detection system for critical infrastructures. *Big Data Research*, 3:10 – 23, 2016. Special Issue on Big Data from Networking Perspective.

[32] L. Dai, B. Wang, Y. Yuan, S. Han, I. Chih-lin, and Z. Wang. Non-orthogonal multiple access for 5G: solutions, challenges, opportunities, and future research trends. *IEEE Communications Magazine*, 53(9):74–81, 2015.

[33] Z. Yang, Z. Ding, P. Fan, and N. Al-Dhahir. A General Power Allocation Scheme to Guarantee Quality of Service in Downlink and Uplink NOMA Systems. *IEEE Transactions on Wireless Communications*, 15(11):7244–7257, 2016.

[34] I. Alimi and O. Aboderin. Adaptive Interference Reduction in the Mobile Communication Systems. volume 1, pages 12–19, 2015.

[35] Y. Gu, Z. Chang, M. Pan, *L. song*, and Z. Han. Joint Radio and Computational Resource Allocation in IoT Fog Computing. *IEEE Transactions on Vehicular Technology*, 67(8):7475–7484, 2018.

[36] Y. Lan, X. Wang, D. Wang, Z. Liu, and Y. Zhang. Task Caching, Offloading, and Resource Allocation in D2D-Aided Fog Computing Networks. *IEEE Access*, 7:104876–104891, 2019.

[37] S. Tong, Y. Liu, M. Cheriet, M. Kadoch, and B. Shen. UCAA: User-Centric User Association and Resource Allocation in Fog Computing Networks. *IEEE Access*, 8:10671–10685, 2020.

[38] T. Huang, W. Yang, J. Wu, J. Ma, X. Zhang, and D. Zhang. A Survey on Green 6G Network: Architecture and Technologies. *IEEE Access*, 7:175758–175768, 2019.

[39] B. Zong, C. Fan, X. Wang, X. Duan, B. Wang, and J. Wang. 6G Technologies: Key Drivers, Core Requirements, System Architectures, and Enabling Technologies. *IEEE Vehicular Technology Magazine*, 14(3):18–27, Sep. 2019.

[40] W. Xia, X. Zhang, G. Zheng, J. Zhang, S. Jin, and H. Zhu. The interplay between artificial intelligence and fog radio access networks. *China Communications*, 17(8):1–13, 2020.

[41] J. Liu, B. Zhao, Q. Xin, J. Su, and W. Ou. DRL-ER: An Intelligent Energy-aware Routing Protocol with Guaranteed Delay Bounds in Satellite Mega-constellations. *IEEE Transactions on Network Science and Engineering*, pages 1–1, 2020.

[42] K. M. Sim. Agent-Based Approaches for Intelligent Intercloud Resource Allocation. *IEEE Transactions on Cloud Computing*, 7(2):442–455, April 2019.

[43] J. Park, S. Samarakoon, M. Bennis, and M. Debbah. Wireless Network Intelligence at the Edge. *Proceedings of the IEEE*, 107(11):2204–2239, Nov 2019.

[44] S. Zhang, J. Liu, H. Guo, M. Qi, and N. Kato. Envisioning Device-to-Device Communications in 6G. *IEEE Network*, 34(3):86–91, May 2020.

[45] K. M. Sim. Agent-based Cloud commerce. In *2009 IEEE International Conference on Industrial Engineering and Engineering Management*, pages 717–721, Dec 2009.

Chapter 2

Low-Latency Strategies for Service Migration in Fog Computing Enabled Cellular Networks

Jun Li, Xiaoman Shen, Lei Chen and Jiajia Chen

Abstract

This chapter presents a fog computing enabled cellular network (FeCN), in which the high user-mobility feature brings critical challenges for service continuity under stringent service requirements. Service migration is promising to fulfill the service continuity during mobility. However, service migration cannot be completed immediately and may lead to situations where the user-experience degrades. For this, a quality-of-service aware service migration strategy is proposed. The method is based on existing handover procedures with newly introduced distributed fog computing resource management scheme to minimize the potential negative effects induced by service migration. The performance of the proposed schemes is evaluated by a case study, where realistic vehicular mobility pattern in the metropolitan network of Luxembourg is used. Results show that low end-to-end latency for vehicular communication can be achieved. During service migration, both the traffic generated by migration and the other traffic (e.g., control information, video) are transmitted via mobile backhaul networks. To balance the performance of the two kinds of traffic, a delay-aware bandwidth slicing scheme is proposed. Simulation results show that, with the proposed method, migration data can be transmitted successfully within a required time threshold, while the latency and jitter for nonmigration traffic with different priorities can be reduced significantly.

Keywords: fog computing, mobile backhaul, low-latency service migration, bandwidth slicing, distributed fog resource management, vehicular communications

1. Introduction

The future cellular network is envisioned to support a variety of emerging mission-critical services such as industrial automation, cloud robotics, and safety-critical vehicular communications [1]. These mission-critical services usually have stringent requirements on latency, jitter, and reliability. In general, the required end-to-end latency is in the order of millisecond, while the probability that this requirement is met is expected to be as high as 99.999%. For example, the communication latency between sensors and control nodes for industrial automation has to be lower than 0.5 milliseconds, while that for virtual and augmented reality has to be lower than 5 milliseconds [1]. As an integral part of the cellular network, the transport network, referred to as the segment in charge of the backhaul of radio base stations and/or the fronthaul of remote radio unit, plays an especially important role to meet such a stringent requirement on latency.

The latency in transport networks can be reduced by moving the computing, storage, control, and network functions to the edge of the network, referred to as fog computing or edge computing, instead of performing all the functions in remote data centers. Fog computing is a new paradigm that can be integrated with the existing cellular networks (e.g., aggression points, base stations) to provide ultra-low-latency communication for time-critical services [2]. Thus, end users can access the applications (e.g., remote driving) hosted in fog nodes with low transport latency.

A fog node can be a terminal or a stand-alone node, which can be co-located with the existing cellular network infrastructure, such as router, gateway, aggregation points, and base stations (BS) [3]. Among them, BS (e.g., LTE evolved Nodes B) is a promising segment that can be integrated with fog nodes, which forms BS-Fog, giving rise to a new concept of fog enabled cellular networks (FeCNs). Such FeCN can be a promising candidate to support real-time services (e.g., real-time vehicular services) due to the ubiquitous access to radio access network (RAN) infrastructure as well as low communication delay enabled by fog computing. **Figure 1** illustrates the overall FeCN architecture.

In the FeCN, the BSs are responsible for providing network functions (e.g., handovers), whereas computational resources (e.g., computing and storage capability) can be provided by the fog nodes locally. One BS-Fog can cooperate with other BS-Fogs or cloud to allocate tasks dynamically. We design the FeCN with minimal changes on the current network architecture and reuse the existing interfaces. The S1 interface, which has been defined as the interface between the BS and the evolved packet core in LTE networks, is considered to realize the communications between the BS-Fog and cloud for the FeCN, while the interface X2, which has been defined as the interface between two BSs, is considered to support the communications among the BS-Fogs.

In the FeCN, it is possible to provide computing and storage capability closer to end users to support time-critical services. However, there are several challenges remaining to be addressed. Firstly, a fog node may be overloaded due to limited computing resources. Secondly, for high-mobility end users, the limited coverage of

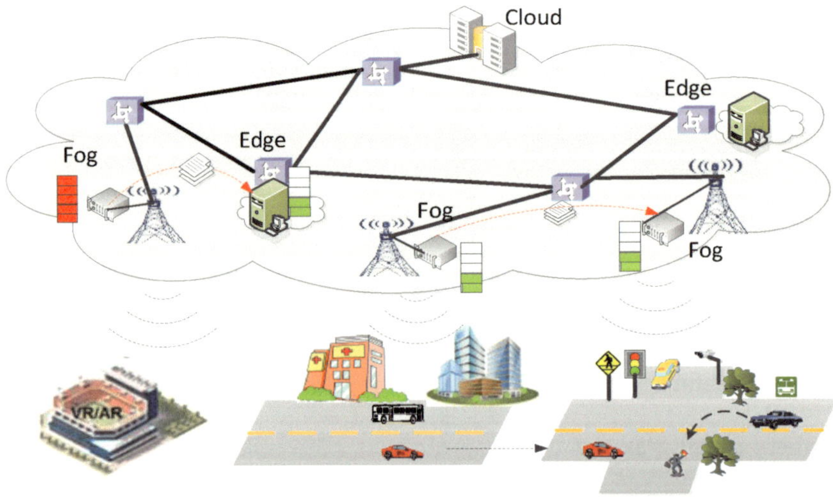

Figure 1.
Service migration in fog computing enabled cellular networks.

a single fog node may result in performance degradation in terms of latency when the user is moving far away from the serving node. In both cases, services need to be migrated to satisfy the quality-of-service (QoS) requirement. Service migration [4], which is referred to as relocating services from one fog node to another, has been proposed to deal with such challenges. As shown in **Figure 1**, once a single fog node is overloaded, service migration is triggered to offload this fog node to other fog nodes or edge servers. Besides, when users move from the area covered by one fog node to another, time-critical services are required to be migrated accordingly in order to follow the users' movement and maintain the service continuity with satisfying stringent latency requirements for these services. Since service migration may take time, which may result in service interruption, it needs to be handled carefully with consideration of the service requirements of differentiated traffic and the available network resources.

This chapter describes mechanisms and algorithms to deal with service migration in the FeCN. In Section 2, a QoS-aware service migration strategy is proposed. The strategy is based on the existing handover procedures, and the performance is studied with connected vehicle use cases. Following the proposed service migration strategy, in Section 3, a distributed fog computing resource management scheme is proposed to deal with limited computational resources at fog nodes. The scheme considers services with differentiated priority level, and in case of resource shortage, low-priority services may be migrated to other fog nodes to guarantee sufficient computation resources for the migrated high-priority services. The performance of the proposed schemes is evaluated by a case study, where realistic vehicle mobility pattern in the metropolitan network scenario of Luxembourg is used to reflect the real-world environment. Results show that low end-to-end latency (e.g., 10 ms) for vehicular communication can be achieved with typical vehicle mobility.

During service migration, both the traffic generated by migration (referred to as migration traffic) and other traffic (referred to as non-migration traffic, e.g., control information, video) are transmitted via mobile backhaul networks. To balance the performance of the two kinds of traffic, in Section 4, a delay-aware bandwidth slicing scheme is proposed in PON-based mobile backhaul networks. The proposed slicing scheme on one hand tries to guarantee the performance of the migration traffic, while on the other hand trying to minimize the negative impact on non-migration traffic. Simulation results show that, with the proposed method, migration data can be transmitted successfully within a required time threshold, while the latency and jitter for non-migration traffic with different priorities can be reduced significantly.

2. Service migration strategy in FeCN

Paper [5] presents a QoS-aware service migration strategy with connected vehicles as a use case to represent the high-mobility characteristic. In the context of FeCN, a vehicle is traveling while accessing a fog node. To maintain the service continuity, the vehicle needs to continue accessing the fog node and the provisioned running services through backhaul networks. When the vehicle travels away from the serving BS-Fog, the end-to-end (E2E) latency will increase, especially for the case that the vehicle does not have fixed routes. In order to keep the vehicle always accessing the fog services in one hop, the ongoing service can be migrated following the vehicle's trace. One straightforward strategy is to perform migration in combination with the handover procedure. This can guarantee one-hop access where a vehicle can directly access the services at its associated BS-Fog. However, since service migration cannot always be completed immediately, this may lead to a situation where users experience loss of service access. Therefore, frequent service migration is not always a good choice, and service migration needs to consider the QoS requirements of services.

2.1 QoS-aware service migration strategy

In view of the disadvantages of handover-based service migration strategy, we present a QoS-aware service migration strategy which is based on the existing handover procedure and considers the service QoS requirements. The key idea is to minimize the migration overhead while maintaining the E2E latency at an acceptable level under certain QoS requirements. E2E latency is a key QoS metric for real-time vehicular communication. When the E2E latency is at an unacceptable level, the performance of other metrics (e.g., reliability and packet drop) can also get worse. Therefore, in the proposed scheme, we focus on the metric of E2E latency to explain our proposed QoS-aware service migration strategy. The generalization of the proposed scheme to other QoS metrics is straightforward.

Figure 2 illustrates the service migration strategy, and **Figure 3** shows the communication protocol for the proposed QoS-aware service migration scheme. As illustrated, once the QoS requirements cannot be satisfied, the source BS-Fog node will trigger the service migration procedure and sends a Migration Request message that contains the information about the QoS requirements of the affected services to the target BS-Fog. After receiving the Migration Request message, the target BS-Fog will first make a decision whether to accept or not, and then sends back a Migration Request ACK to inform the source BS-Fog of its decision. If the request is agreed, the source Fog-BS will start implementing the migration.

In the proposed scheme, service migration can be achieved by pre-copy technique which is widely used for live virtual machine (VM) migration, as presented in [6]. The migration can be performed in two phases. In the first phase, the transfer of memory pages to the target BS-Fog is completed iteratively without suspending VM. In this phase, the UE still accesses the source BS-Fog (see blue dashed line in **Figure 3**). In the second phase when sufficient memory pages are transferred, the

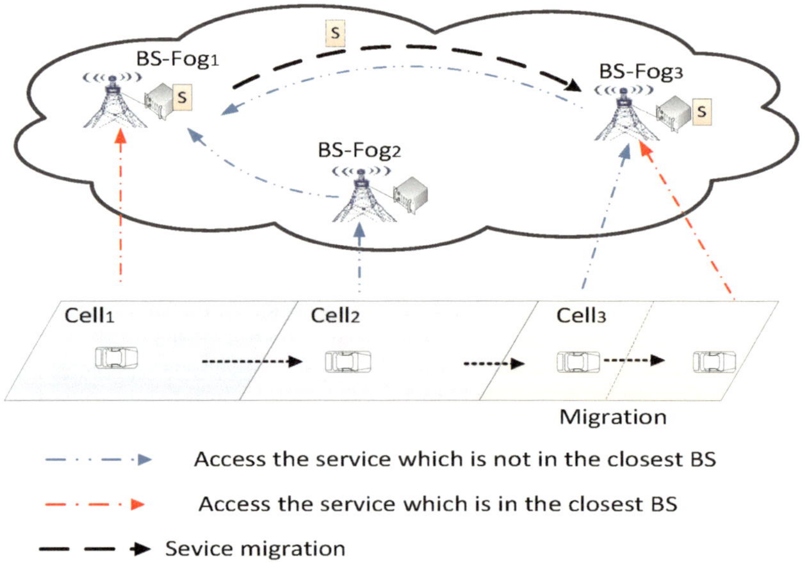

Figure 2.
Illustration of a QoS-aware service migration [5].

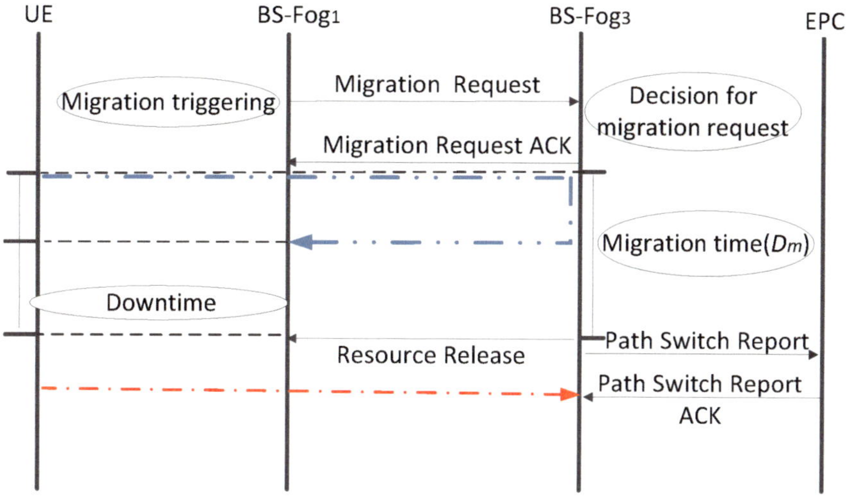

Figure 3.
Communication protocol to support QoS-aware service migration [5].

source BS-Fog will suspend the VM and finish transferring the remaining memory pages to the target BS-Fog. The services cannot be properly accessed during this period when the VM is suspended. Such duration is denoted as downtime. After the migration is completed, the UE can directly access the service in one hop (see red dashed line in **Figure 3**).

2.2 Performance evaluation

In this section, the performance of the proposed migration strategy is evaluated by using simulation. Here, the fog computing resources allocated for migrated services are assumed to be sufficient. This can be realized by designing efficient fog computing resource management schemes. We consider a case study for a small service area by using the realistic mobility pattern for the country of Luxembourg, and BS-Fog entities are evenly distributed over the city. The parameters used in this simulation are described in **Table 1**.

The performance of the proposed delay-aware migration (named Scheme 3) is evaluated in comparison to two benchmarks: no service migration (named Scheme 1) and always service migration (named Scheme 2). The details of Scheme 1 and Scheme 2 are introduced in paper [5]. The handover interruption time and wireless delay cannot be ignored and are not affected by migration strategies. Therefore, it is assumed that the uplink delay in the wireless segment is within 0.5 ms and the handover interruption time is a constant.

The average E2E latency for the three schemes is shown in **Figure 4**. Here, E2E latency consists of wireless access delay, interruption time during the handover, migration time, backhaul delay, and processing and queuing delays at the BS-Fogs. The transmission capacity is denoted as B and refers to the bandwidth allocated to the X2 interface between two Fog-BSs. It can be seen that the E2E latency in all three schemes decreases as B increases, especially for Scheme 1. This is because higher bitrate leads to shorter packet transmission time. Thus, the packet queueing delay can be reduced, resulting in smaller access latency. When B is high enough (e.g., B = 240 Mbps in **Figure 4**), the queueing delay is as minor as negligible. Meanwhile, in Scheme 2, the E2E latency is mainly affected by downtime (D_t), during which the ongoing services need to be suspended.

Parameter	Value
Coverage of the country of Luxembourg	155 km^2 [7]
Total number of vehicles	5500 [7]
Vehicle density	35.5 per km^2
Bitrate of traffic generated by the vehicles	(2 Kbps, 10 Mbps)
Size of the applications encapsulated in VMs	(10,100) Mbits
Link speed in upstream and downstream in PONs	10 Gbps
Repeated times of simulations	10
Simulation time	1000 s
Coverage of a single BS-Fog	1 km^2
Handover interruption time	20 ms [8]
Wireless delay	0.5 ms
Vehicle speed	(1, 45) m/s [7]
Processing time at active node	0.2 ms
Number of ONUs in each PON	16
Number of PONs	10
Confidence level	95%

Table 1.
Simulation parameters.

In Scheme 3, the service migration is triggered once the latency exceeds the threshold (e.g., 10 ms in **Figure 4**). Scheme 3 is considered as a tradeoff between Scheme 1 and Scheme 2. When B is low, Scheme 3 performs similar to Scheme 2, that is, frequent migrations are performed in both schemes (see **Figure 5**). As B increases, there are fewer and fewer migrations triggered in Scheme 3. Thus, the E2E latency is less and less affected by downtime. Therefore, Scheme 3 has similar

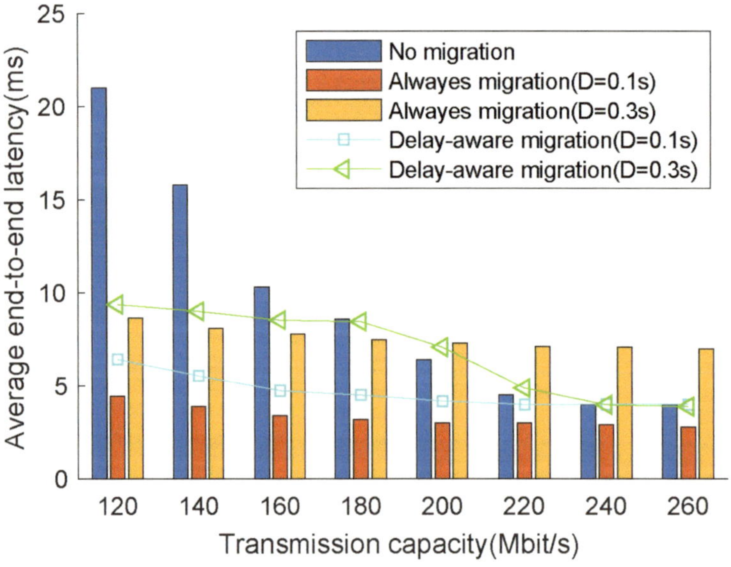

Figure 4.
Transmission capacity (B) in the backhaul versus average end-to-end latency.

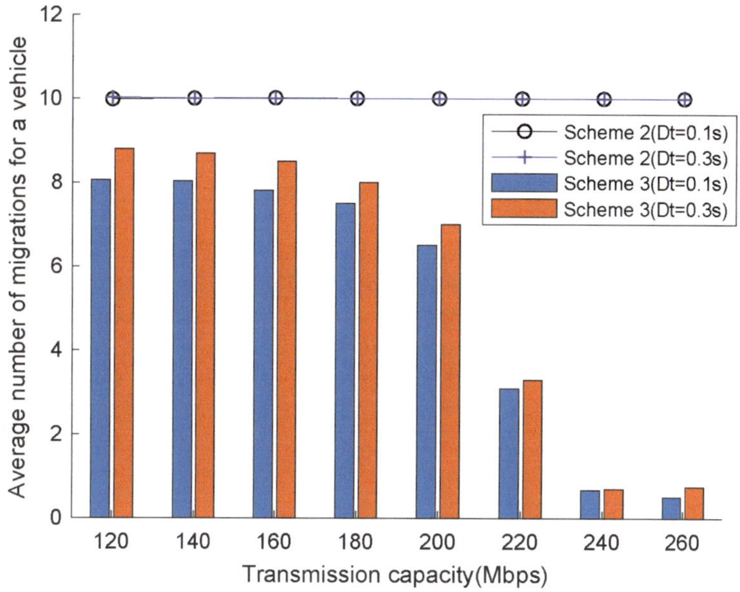

Figure 5.
The average number of migrations for a vehicle as a function of transmission capacity.

performance with Scheme 1 when B is high and performs better than Scheme 2 when downtime is large (e.g., 0.3 s), as shown in **Figure 5**.

3. Distributed fog computing resource management

During the service migration procedure, sufficient computation resource is needed to host the migrated services. Also, provisioning resource for the migrated real-time services needs to be completed as soon as possible to minimize the service interruption. Otherwise, the services have to stay at the source fog node, which may increase the access delay. On the other hand, one single fog node only has limited amount of computation resource, and its load is highly burst due to the mobility of vehicles. The service migration strategy discussed in the previous section assumes sufficient resources for the migrated services, which may not always hold in practice. Thus, it is important to employ efficient resource management scheme, especially in the scenarios with fast mobility and high load.

To guarantee sufficient computation resource for the migrated vehicular services and thus reducing the service migration blocking caused by the lack of computing resources, a distributed fog computing resource management scheme is proposed [9]. Two distributed fog computing resource management schemes, namely, fog resource reservation (FRR) and fog resource reallocation (FRL), have been considered. In FRR scheme, a certain amount of computation resources for vehicular services in each fog node are reserved based on the predicted vehicular traffic load. The performance of this scheme depends on the traffic flow prediction methods. Overestimating leads to low resource utilization, while underestimating significantly decreases one-hop access probability for high-priority (HP) vehicular services (e.g., remote driving, pre-crash sensing warning). For FRL, the key idea is to release part of fog resources used for low-priority (LP) services (e.g., online game, navigation, sign notification) by suspending those services and reallocate them to HP services. However, in such a scheme, the one-hop access probability for

LP services may be affected, especially when traffic load is high. In fact, not all the LP services (e.g., online game) need to have one-hop latency requirement and local awareness. Therefore, such services can be placed in its neighboring fog nodes with low load.

3.1 Distributed computing resource management

As introduced above, in both FRR and FRL, each BS-Fog node manages its resource independently without cooperating each other. Once it is overload, the one-hop access probability for LP services may be affected. Considering that some LP services also have one-hop latency requirement, we proposed an online resource management (ORM) scheme, in which fog nodes can cooperate to allocate resources for both HP and LP services [10]. In comparison to FRL, LP services are not suspended, but instead, they are migrated to other neighboring BS-Fogs. This can guarantee the resource requirement of the to-be-migrated HP services as well as LP service continuity. The details are as follows.

When a vehicle moves from one BS-Fog's coverage area to another, both handover and migration of ongoing services need to be handled. As shown in **Figure 6**, once the handover is triggered, the source Fog-BS sends a Migration Request message to the target BS-Fog, which includes the information of the requested resources. After receiving this message, the target Fog-BS will make a decision

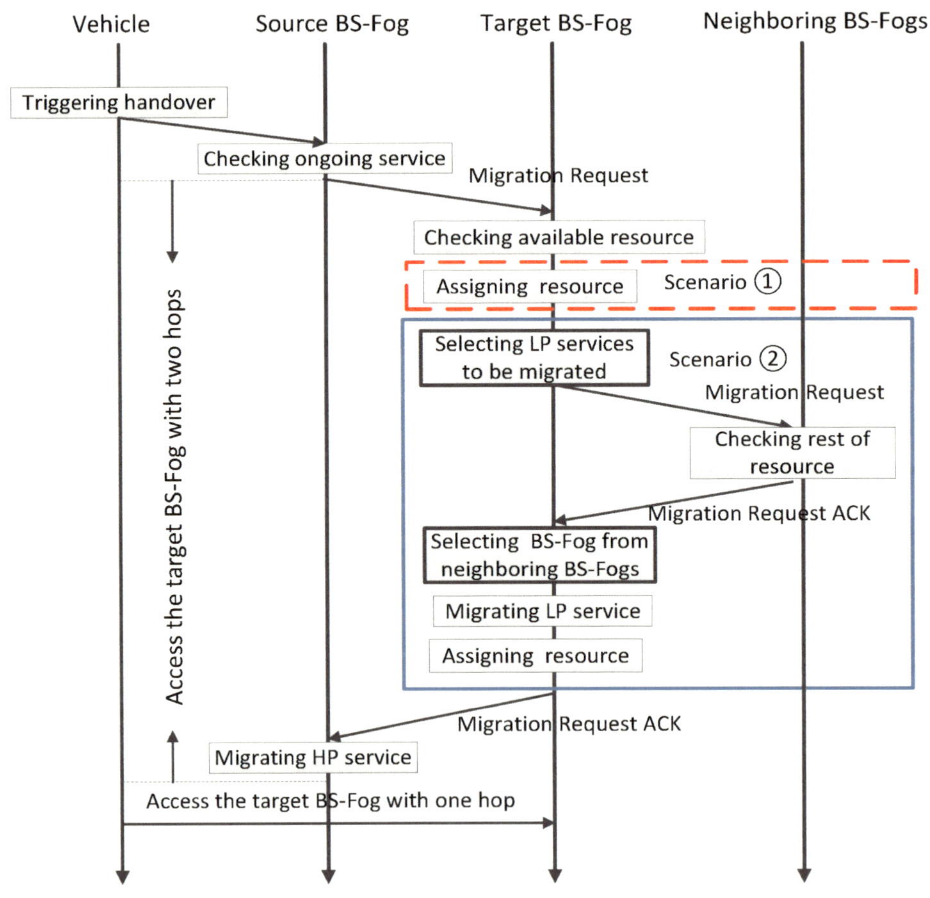

Figure 6.
Illustration of resource management in service migration procedure [10].

whether to accept the request or not according to the resource management strategy.

In this part, we consider two scenarios used for the resource management strategy. In the first scenario, there is sufficient available resource in the target BS-Fog, the request is approved, and the corresponding resource is assigned to the migrated HP services (see Scenario ① in red dotted box in **Figure 6**).

In the second scenario, the target BS-Fog does not have enough available resource. In such a case, the target BS-Fog selects ongoing LP services and migrates them to its neighboring BS-Fogs. After that, the resources of the selected LP services are released for HP services. As shown by Scenario ② in blue dashed box in **Figure 6**, the procedure is divided into two phases. In the first phase, LP services to be migrated are selected according to service selection strategies, while in the second phase, the target BS-Fog selects one of its neighboring BS-Fogs to host the migrated LP services. For such a purpose, the target BS-Fog firstly broadcasts a Migration Request message to its neighboring BS-Fogs that have direct communication with the target BS-Fog via X2 interface. Once the neighboring BS-Fogs receive the Migration Request messages, they will check their available resource and then send Migration Request ACK to the target BS-Fog. A final decision on the selection of neighboring BS-Fog is made by the target BS-Fog immediately. When the selected LP services complete the migration and release their resources, the target BS-Fog sends a Migration Request ACK to the source BS-Fog for HP service migration. If there are no available LP services or neighboring BS-Fogs, the to-be-migrated HP service has to run at the source BS-Fog and the vehicle has to access the service with more than one hop (i.e., the hop(s) between two neighboring BS-Fogs have to be counted for access), which may obviously result in a higher access delay. The selection strategies of LP services and neighboring BS-Fogs are discussed in the following section.

3.1.1 Low-priority service selection algorithm

In the proposed scheme, the selected LP services are migrated from the target BS-Fog to the selected neighboring BS-Fog, which will be accessed via backhaul networks. In such a procedure, a certain amount of backhaul bandwidth is needed for service migration and running the selected LP services. To minimize the required bandwidth, we propose a LP service selection algorithm to minimize the communication cost. The amount of the required bandwidth resources for each LP service is firstly counted. Then, the LP services whose available computing resources are larger than the requested amount are selected. Among these services, the one with the lowest communication cost is finally selected. If there is no service that satisfies the requirement alone, more than one service can be selected. In order to avoid ping-pong effect in LP service migration, LP services are only allowed to be migrated once.

3.1.2 Neighboring fog-BS selection algorithm

Once the LP services to be migrated are decided, neighboring BS-Fogs that will host the migrating services need to be decided. The decision needs to consider the QoS of those services since after migration, the services will be hosted at the selected neighboring BS-Fog and accessed with two hops, which results in extra backhaul delay. As LP services are only allowed to be migrated once, the access delay for the migrated LP services consists of radio access delay and backhaul delay. According to the delay requirement of the selected LP service, the budget of backhaul delay can be calculated. The transmission delay between the target BS-Fog and its neighboring BS-Fog should thus be smaller than the budget of backhaul delay. The key idea of the proposed algorithm is to select the neighboring BS-Fog with the most available resources under the acceptable backhaul delay of the to-be-migrated LP service.

3.2 Performance evaluation

In this section, the performance of the proposed scheme is investigated through simulation. The realistic mobility pattern for the city of Luxembourg is used. **Figure 7** shows the vehicular traffic profile in Luxembourg, which varies in time over a day. Also, the vehicular traffic is spatially diverse. For example, the inserted chart (a) in **Figure 7** shows the numbers of vehicles in each coverage area of BS-Fogs at 8:00 am, while the inserted chart (b) in **Figure 7** shows the numbers of vehicles at 12:00 pm. The Y-axis shows the number of vehicles, while the X-axis is the series number of BS-Fog.

Each vehicle is assumed to only require 1 HP service (i.e., safety-related service). The data traffic distribution is proportional to the vehicular traffic. Without loss of generality, we assume the total service request arrival rate for the BS-Fog network is in the range from 20 to 100 (per second). The arriving requests consist of 30% of HP and 70% of LP services. The HP service arrival rate is distributed among the BS-Fogs according to the traffic profile at 8:00 am (see **Figure 7(a)**), while the LP service arrival rate is distributed among BS-Fogs evenly. Both HP and LP services arrive according to Poisson Procedure. The parameters are shown in **Table 2**.

The performance of the proposed scheme is investigated in terms of access probability for HP services and service unavailability for LP services. Here, one-hop access probability is defined as the ratio of the one-hop service access duration to the total holding time. Similarly, service unavailability is defined as the ratio of the time when service is not available to the total holding time. Here, the services may be unavailable due to the lack of resources in the current BS-Fog and its neighbors, as well as due to the interruption during service migration. As discussed, to increase one-hop access probability while reducing service unavailability, migration time for both HP and LP services should be minimized, which is related to the transmission time in the mobile backhaul network. Transmission capacity B becomes the main factor that affects the delay performance.

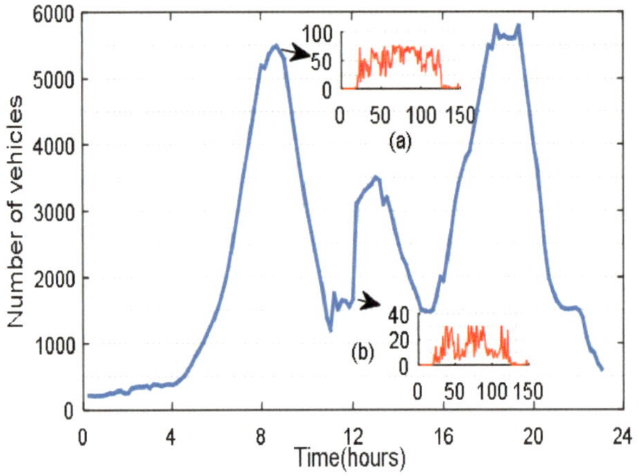

Figure 7.
Vehicular traffic profile in Luxembourg [10].

Parameter	Value
Total number of computing units in one fog	400
Number of BS-Fogs in the network	100
Number of computing units for each HP service	3
Number of computing units for each LP service	(2, 6)
Average serving time in one fog for HP service (second)	90
Standard deviation of the serving time for HP service (second)	10
Average serving time in one fog for LP service (second)	120
Budget of backhaul delay for LP services (millisecond)	(5, 10)
Data rate generated by end users (bps)	(2 K, 10 M) [11]
Amount of data encapsulated in application VMs (Mbits)	(10, 100)
Downtime in live VM migration (millisecond)	20
Confidence level	95%

Table 2.
Simulation parameters [10].

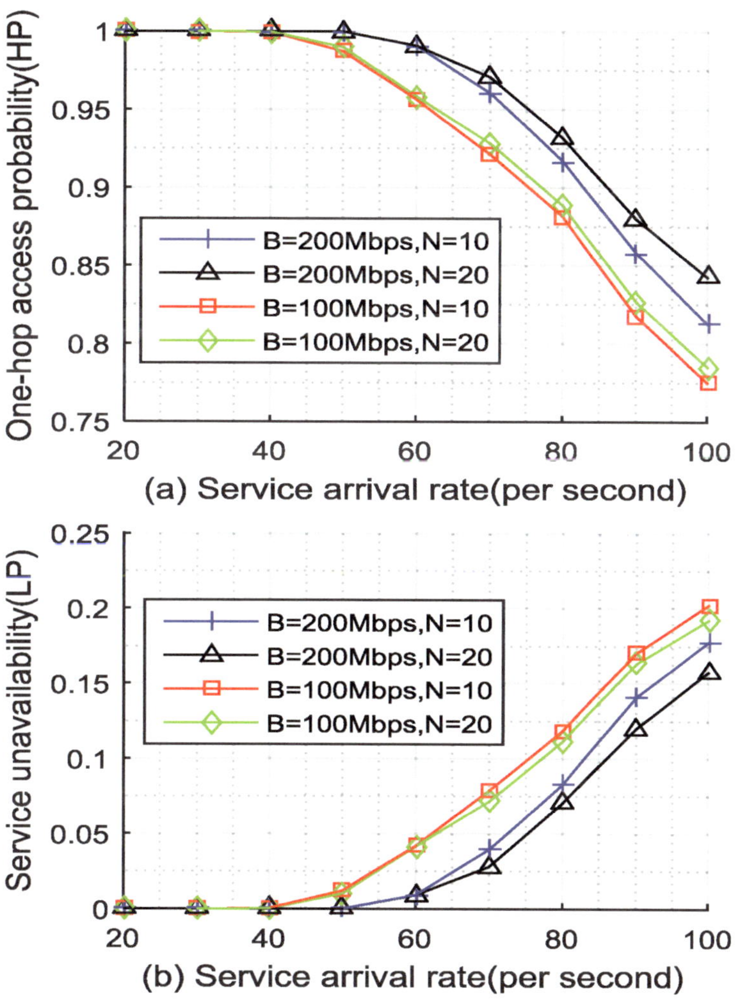

Figure 8.
One-hop access probability and service unavailability versus service arrival rate.

Figure 8(a) shows that one-hop access probability for HP services versus service arrival rate. As expected, due to the fact that the migration delay for both HP and LP services can be reduced by enlarging backhaul capacity, larger transmission capacity (*B*) leads to higher one-hop access probability, which benefits the reduction of migration time. As also shown, when B is large (e.g., B = 200Mbps), a higher number of neighbors (*N*) lead to a better one-hop access probability, while when B is small, the increase of N has little impact on the one-hop access probability for HP services. This is because with a smaller B, the backhaul delay between the target and the neighboring BS-Fogs is higher, and the number of neighboring BS-fogs that satisfy the latency requirement decreases, even when N is high. **Figure 8(b)** shows LP service unavailability as a function of service arrival rate. Similarly, increasing backhaul capacity B leads to a lower migration delay and thus a reduced service unavailability.

We further compare the performance of the proposed ORM scheme with two benchmarks. The first benchmark is based on the principle of first come first served (FCFS), in which HP and LP services are treated equally. The second benchmark is FRR where a certain amount of resource is reserved for HP. **Figure 9(a)** shows that, in comparison to FCFS and FRR, the one-hop access probability of HP services for ORM is higher when B is large (e.g., B = 200 Mbps). When B decreases to 100 Mps,

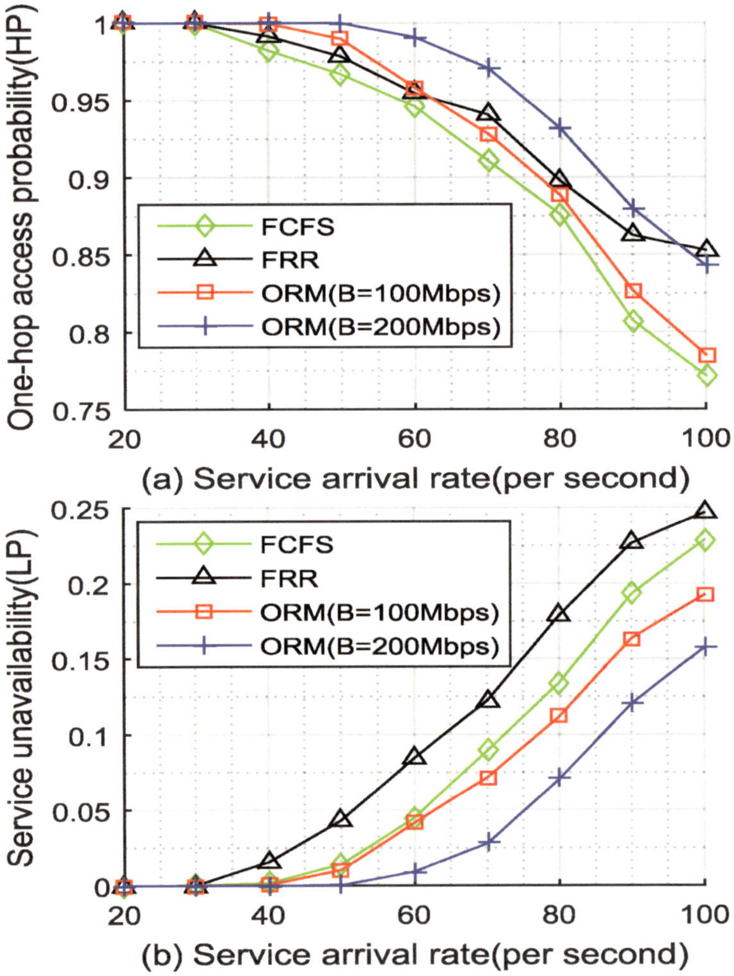

Figure 9.
One-hop access probability and service unavailability versus service arrival rate.

the one-hop access probability for ORM shows different results based on the service arrival rate. When the service arrival rate is below 60 arrivals per second, it is higher than that for both FCFS and FRR, while when the service arrival rate increases above 60, the one-hop access probability remains higher than that for FCFS but lower than that for FRR. The reason is that more LP services need to be migrated to the neighboring BS-Fogs when the service traffic arrival rate increases, which results in high migration traffic, thus increasing the migration delay.

Figure 9(b) shows that the service unavailability for LP services increases with service arrival rate for all methods, and that of ORM is shown to be the lowest in comparison with the two benchmarks. As also can be seen, transmission capacity B has impacts on the service unavailability for LP services. With a higher $B = 200$ Mbps, ORM demonstrates significant lower LP service unavailability at all service arrival rates, that is because a larger B results in a shorter time used for migrating LP services; thus service unavailability can be reduced. When B is decreased to be at 100 Mbps, service arrival rate is shown to affect the performance of ORM. With a service rate under 60 arrivals per second, ORM has very similar performance regarding LP service unavailability in comparison with that for FCFS. When the service rate is above 60, ORM demonstrates the advantages over FCFS with a lower LP service unavailability. The reason is that, in the case of FCFS, the migration delay is mainly introduced by the waiting time for the available resources. Since a larger arrival rate leads to a longer waiting time, the performance of FCFS degrades. As another general observation, though ORM demonstrates better performance, the migration delay is shown as an important factor that affects the performance in terms of one-hop access probability and service unavailability, indicating that migration delay needs to be properly handled.

4. Bandwidth slicing in mobile backhaul networks

In previous sections, service migration strategy and fog computing resource management scheme have been investigated to support real-time vehicular services. As indicated, mobile backhaul capacity is a main factor that affects the performance of the service migration schemes. Regarding this matter, passive optical network (PON) based mobile backhaul network can be considered to support FeCN duc to its high capacity.

In PON-based mobile backhaul network supporting the FeCN, the BS-Fogs are integrated with optical network units (ONUs) through high-speed Ethernet interface, shown in **Figure 10**. The traffic generated by service migration, named migration traffic, is transmitted together with the non-migration traffic. On the one hand, the size of the data generated by service migration can be up to hundreds of MBytes [12]; thus, such migration traffic can be fragmented into Ethernet frames at ONUs and should be carefully handled by the optical line terminal (OLT) that is located at the central office. On the other hand, service migration is usually deadline-driven and has to be handled within a certain time limit. We define migration delay as the time duration from the moment when a migration is initiated until the affected service is successfully transferred to the target fog node. In order to minimize the service interruption, the migration delay should be lower than a pre-defined time threshold, which is usually specified in the QoS requirements and in a magnitude of seconds [13]. The non-migration traffic includes data generated by multi-type applications, which usually have different QoS requirements but less stringent in terms of latency, packet loss ratio, etc. These different types of the non-migration traffic can be queued independently and scheduled with different priorities according to medium access control (MAC) protocol in Ethernet PON [14].

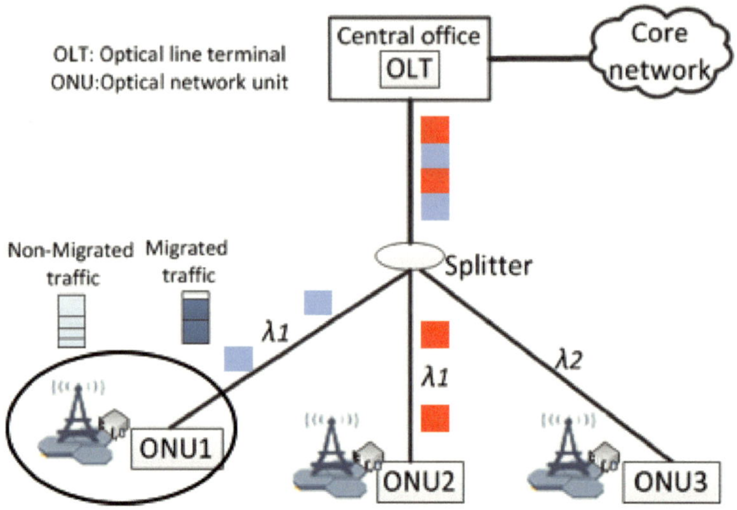

Figure 10.
PON-based mobile backhaul for the FeCN.

Likewise, the migration traffic can be also queued based on their deadline requirements.

Time and wavelength division multiplexing (TWDM-PON) has been regarded as a promising candidate for next-generation PON 2 (NG-PON 2), where dynamic bandwidth allocation (DBA) mechanisms are performed on each wavelength for efficient channel sharing [15]. In a classical DBA algorithm, migration traffic and non-migration traffic are scheduled with no distinction. Each ONU reports to the OLT the amount of data that needs to be transmitted in the next cycle and then receives a grant message. According to the information contained in the grant message, including the allocated time slots and polling cycle for transmission, each ONU transmits data based on the principle of FCFS without considering the traffic priorities. Once a service migration occurs, a large volume of migration data arrives, and more than one polling cycle may be needed for the transmission. In such a case, the non-migration data that arrives after the migration data has to wait before being transmitted, thus experiencing a long queuing delay, leading to high latency and jitter. One way to deal with this is to assign higher priority to non-migration traffic that arrives after the migration traffic based upon the existing QoS management mechanism in Ethernet PON. In such a case, the delay for the non-migration traffic can be reduced significantly. However, short delay for transmitting the migration traffic that comes after may not be guaranteed, particularly when the load of non-migration traffic is high.

For balancing the transmission of migration traffic and non-migration traffic, we propose a dynamic bandwidth slicing (DBS) scheme with a bandwidth slicing mechanism and a tailored delay-aware resource allocation algorithm. We present the DBS in the following part together with simulation results.

4.1 Bandwidth slicing mechanism

A bandwidth slicing scheme for service migration in PON-based mobile backhaul networks is proposed [16]. In the scheme, the cycle time can be cut into several slices dynamically, which are provisioned to different kinds of traffic (i.e., migration traffic and non-migration traffic with different priorities). Such a mechanism is based on the report-grant mechanism, as introduced in the previous section. In each polling cycle, request messages are first sent from each ONU to the OLT containing the information about their data size and delay requirements. According to such information, the polling tables (see **Figure 10**) for both non-migration data and migration data are then updated. Once service migration occurs, the lengths of the slices for both migration and non-migration traffic can be calculated by the resource management controller located at the OLT with the bandwidth allocation algorithms and the information contained in the polling table.

Figure 11 illustrates in more detail the proposed bandwidth slicing mechanism. Following similar principles of the considered DBA algorithms, in each polling cycle, the time slots in Slice 1 and Slice 2 are allocated to the non-migration traffic and the migration data, respectively. As mentioned, the lengths of the slices are decided dynamically based on the traffic and resource allocation algorithms. In the case where there is no migration data to be transmitted, the proposed mechanism performs in the same way as the classical DBA mechanism with a FCFS fashion. Note that the same principle as the proposed mechanism applies to both the upstream and downstream.

4.2 Delay-aware resource allocation algorithm

Following the proposed bandwidth slicing mechanism, a tailored delay-aware resource allocation algorithm is proposed with the aim to transmit migration traffic within the required deadline by cutting the large-size migration traffic into small pieces and transmitting them at each polling cycle. Such an algorithm is implemented at the end of Slice 1 in each polling cycle. First, the OLT acquires the information of the amount of the non-migration traffic that ONUs need to send in the next polling cycle and their priorities contained in the Report messages. If service migration occurs, such messages also contain the information of the sizes and deadline of the migration data that ONUs need to send. Then, the length of the slice for the migration traffic (S^m) can be calculated by OLT, and the migration mechanism will be triggered. In such a case, as illustrated in **Figure 11** the bandwidth slicing mechanism, the polling cycle will be divided into two slices for non-migration traffic (Slice 1) and migration traffic (Slice 2), respectively. The time slots in Slice 1 are allocated to ONUs for non-migration traffic according to their priority level, while the time slots in Slice 2 are allocated to the migration traffic according to the ascending order of the deadline for finishing migration. Here time slot allocation is purely based on incoming traffic, and with high migration traffic load, it can be expected that the time slots are monopolized. To avoid such situation,

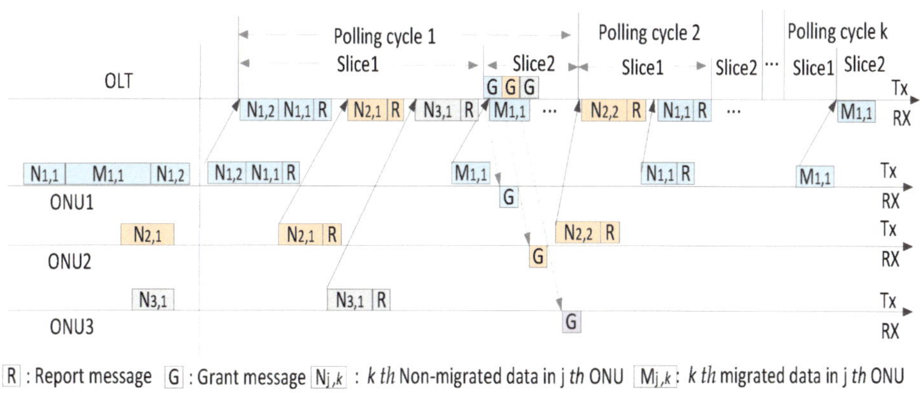

Figure 11.
Illustration of the proposed bandwidth slicing mechanism.

Symbol	Explanation
$R_i^{j,k}$	Required time slot for the remaining of the kth migration traffic from the ONUj in the ith polling cycle
$D^{j,k}$	Deadline for the kth migration traffic sent by ONUj
$T_i^{j,k}$	Remaining time for transmitting the kth migration traffic sent by ONUj, which starts from the beginning of the ith polling cycle to its deadline
K_j	Number of the migration tasks in the ONUj
B^R	Transmission time for sending each report and grant massage
B^G	Guard time with a fixed value for the slice of the migration traffic in the ith polling cycle
$B_i^{j,l}$	Length of the requested time slots for the non-migration traffic with the lth priority at the in the ith polling cycle
H_j	Number of priority levels for the non-migration traffic at the ONUj

Table 3.
Explanation of symbols.

a hard threshold θ is introduced as the percentage of the total time slots that can be allocated to migration traffic within each polling cycle.

In the proposed algorithm, the calculation of the lengths of slices in each polling cycle plays a very important role and is described in more detail in the following part. The symbols used are explained in **Table 3**. In each polling cycle (e.g., i^{th} polling cycle), the required time slots ($G_i^{j,k}$) for the kth migration traffic from the ONUj can be calculated by

$$G_i^{j,k} = R_i^{j,k} / \left\lceil T_i^{j,k} / W_{max} \right\rceil \quad (1)$$

In the proposed resource allocation algorithm, the length of the polling cycle (W) varies dynamically with the traffic load. Thus, when calculating the required time slots in the current polling cycle, the maximum polling cycle (W_{max}) is used to guarantee that the transmission of the whole migration traffic can be finished before the deadline. Here, the time unit (μs) is used to represent the length of the time slots and polling cycles. Then, the total length of the time slots granted for the migration traffic ($TG_i^{j,k}$) can be calculated by

$$TG_i^{j,k} = \sum_{j=1}^{N} \sum_{k=1}^{K_j} G_i^{j,k} \quad (2)$$

To guarantee the fairness between the migration and non-migration traffic, the length of the granted time slots cannot exceed the maximum allowed length of the time slots in this polling cycle, which can be calculated by.

$$R_i^m = \left(W_{max} - N \times \left(B^R + B^G\right)\right) \times \theta \quad (3)$$

The maximum length of the allocated time slots for the migration traffic is set by the threshold (θ) for the slice of the migration traffic ($\theta \in [0, 1]$). Thus, the time slots granted for the migration traffic in the ith polling cycle can be calculated by

$$TG_i^{j,k} = \begin{cases} TG_i^{j,k}, & TG_i^{j,k} < R_i^m \\ R_i^m, & TG_i^{j,k} \geq R_i^m \end{cases} \tag{4}$$

In the ith polling cycle, the length of the slice for the migration traffic (S_i^m) equals to the total length of the granted time slots ($TG_i^{j,k}$). Then, for the non-migration traffic (C_i^t) with different priorities, the total length of the granted time slots can be calculated by

$$C_i^t = \sum_{j=1}^{N} \sum_{l=1}^{H_j} B_i^{j,l} \tag{5}$$

For the non-migration traffic, the maximum available time slot (C_i^a) can be calculated by

$$C_i^a = W_{max} - S_i^m - N \times \left(B^R + B^G\right) \tag{6}$$

Then, the granted time slots can be calculated by

$$C_i^t = \begin{cases} C_i^t, & C_i^t < C_i^a \\ C_i^a, & C_i^t \geq C_i^a \end{cases} \tag{7}$$

Similarly, in the ith polling cycle, the length of the slice for the non-migration traffic (S_i^n) equals the total length of the granted time slots (C_i^t).

4.3 Performance evaluation

The performance of the proposed algorithm has been investigated through simulation and is also further compared with two benchmarks that are based on the conventional DBA algorithms [15]. In Benchmark1, the migration traffic and non-migration traffic follow FCFS, while in Benchmark2 a higher priority is given to the non-migration traffic. Besides, in both benchmarks, the non-migration traffic is assumed with two priority levels (e.g., low and high). **Table 4** summarizes the main parameters.

As mentioned, a threshold θ is introduced to regulate the allocation of time slots within each polling cycle. It has been shown that with θ set to 1, 85% of the time slots

Parameter	Value
Number of ONUs in a PON	8
Propagation delay in the optical links	5 μs/km
Packet size of Ethernet frame (bytes)	(64, 1518)
Guard time between two consecutive time slots	1 μs
Buffer size (Mbytes)	100
Amount of data encapsulated in application VMs (Mbits)	(10, 50)
Deadline for the migration data (second)	(1, 5)
Confidence level	95%

Table 4.
Simulation parameters [16].

in the overall polling cycle are allocated to the migration traffic at load = 0.9. In the following simulation, different values have been chosen to illustrate the impacts.

Figure 12 illustrates the migration success probability (MSP) versus traffic load. Here, MSP is defined as the ratio of the amount of services migrated before the required deadline over the total amount of services that are migrated. As shown, MSP decreases with increasing traffic load in Benchmark2 and DBS with different thresholds. At a lower traffic load (e.g., less than 0.4), all three schemes achieve high MSP, while when the traffic load is above 0.4, MSP starts to decrease. For Benchmark1, MSP shows very minor changes when traffic load increases with almost 1 when traffic load is 0.9. This is because according to the principle of FCFS, large-size migration traffic can be fully transmitted in several cycles once the migration starts. On the other hand, in Benchmark2, the MSP for the migration traffic decreases sharply due to the fact that non-migration traffic is prioritized. As shown, MSP is as low as 0.1 when the traffic load is 0.9. For DBS, all the migration tasks can be performed within the time constraints when the traffic load is under 0.5. When the traffic load is higher than 0.6, the MSP performance is mainly affected by the threshold of the allowable time slots that can be used for the migration traffic and increases as the threshold increases. For example, with the threshold set to 0.5, MSP can be up to 0.98 at load of 0.7.

The average E2E latency for the non-migration traffic with high priority is shown in **Figure 13(a)** as a function of load. It can be seen that the proposed scheme and two benchmarks have a similar trend, that is, the average latency increases with traffic load. Among the three schemes, Benchmark1 always has the highest average E2E latency which can be up to 100 ms when traffic load is 0.9. Such large latency may not be accepted for time-critical services (e.g., interactive voice). Compared with Benchmark1, the average E2E latency in Benchmark2 increases more slowly even when the traffic load is high. The reason is that the non-migration traffic in Benchmark2 has high priority to be transmitted. Compared with Benchmark1 and Benchmark2, the average E2E latency for DBS is much lower, which is less than 1 ms when the traffic load is lower than 0.5 and remains to be less than 10 ms even at high traffic load. This is due to the fact that the migration traffic can be transmitted in multiple cycles by partitioning those with large size into multiple smaller pieces; thereby the non-migration traffic that

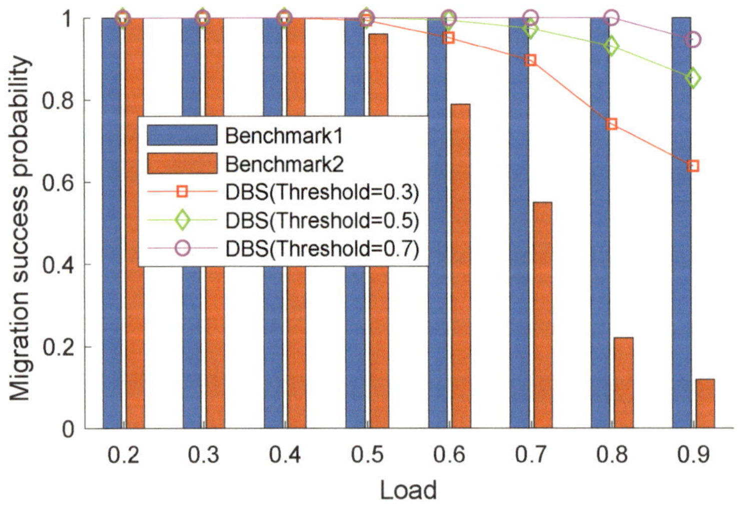

Figure 12.
The migration success probability versus traffic load.

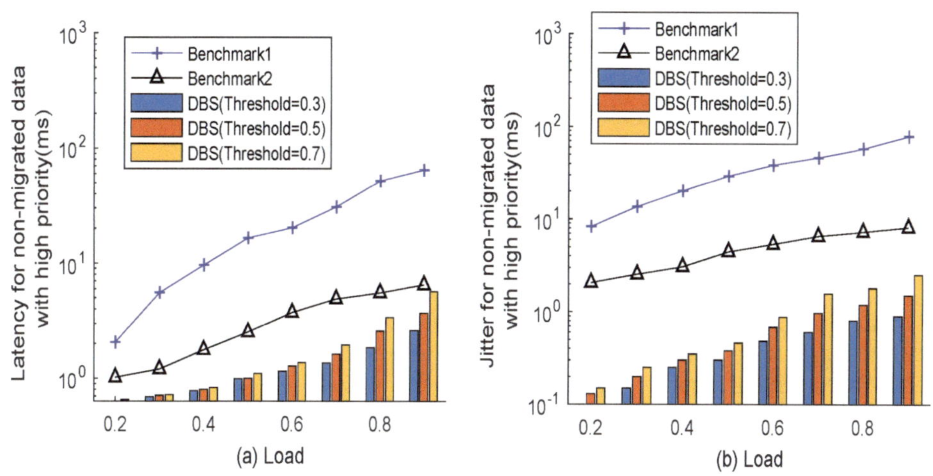

Figure 13.
(a) The average latency and (b) jitter for the non-migration data with high priority.

arrives after or during the transmission of migration traffic does not need to wait too long for transmission. Furthermore, when the threshold increases, the allocated time slots for transmitting the non-migration traffic decreases; thus the average E2E latency increases. Regarding the jitter for the non-migration data with high priority, a similar trend as the average E2E latency can be found, as shown in **Figure 13(b)**.

The average E2E latency of the low-priority non-migration data with different traffic loads is shown in **Figure 14(a)**. Similar to high-priority non-migration traffic, a general trend is that E2E latency increases with the traffic load for all schemes. When the traffic load is low, all kinds of traffic can be assigned with sufficient time slots, while when the traffic load increases, the average E2E latency for the low-priority non-migration traffic increases sharply because of its large queueing delay. More specifically, Benchmark1 has the highest average E2E latency among the three schemes. Compared with Benchmark1, the average E2E latency in Benchmark2 is much lower. And when traffic load is low (e.g., less than 0.6), the average latency of the low-priority non-migration traffic in DBS is the lowest, which is smaller than 2 ms. However, the latency of DBS increases quickly with larger thresholds

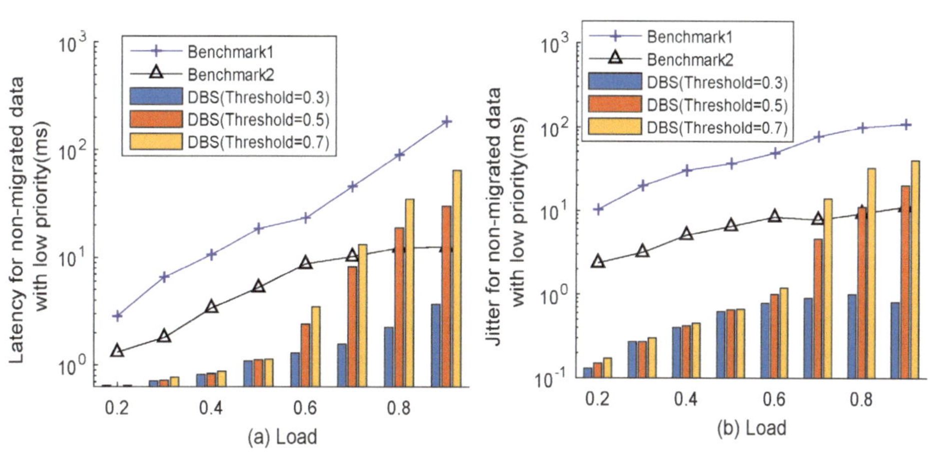

Figure 14.
The average (a) latency and (b) jitter for the non-migration data with low priority.

(e.g., larger than 0.5) and exceeds the level observed for Benchmark2 when traffic load is high (e.g., higher than 0.7). The reason is that the time slots are prioritized for the non-migration traffic with high priority and the migration traffic; thus the low-priority non-migration has to wait. The jitter shows similar trend as the E2E latency, as shown in **Figure 14(b)**.

5. Conclusions

This chapter presents a concept of fog enabled cellular networks (FeCN), where computing, storage, and network functions are provisioned closer to end users in order to improve the service QoS. In addition, to guarantee service continuity and QoS, service migration is introduced to ensure that services always follow the end users through migration from the current fog server to the target one. A QoS-aware service migration strategy based on the existing handover procedures is firstly proposed to balance the benefits and costs of migration. A case study using a realistic vehicle mobility pattern for Luxembourg scenario is carried out through simulation to evaluate the performance of the proposed schemes. Results show that low end-to-end latency (e.g., 10 ms) for vehicular communication can be achieved, while the total number of migrations for each user in the whole journey can be decreased significantly.

To deal with the situation that the target fog node does not have enough resources to support the migrated services, a distributed fog computing resource management scheme is introduced. The scheme purposely selects low-priority (LP) services and migrates those services to carefully selected neighboring fog nodes so that QoS for high-priority (HP) migration services can be served at the target fog node. LP service selection algorithm is proposed to minimize the migration costs for those services, and neighboring fog node selection algorithm is proposed for selecting a fog node that provides enough resources for LP services with also satisfied QoS. Simulation results show that the one-hop access probability for HP services increases significantly, while the service unavailability for LP services can also be well reduced.

During service migration, both the traffic generated by migration and other traffic (e.g., control information, video) are transmitted via mobile backhaul networks. To balance the performance of the two kinds of traffic, we propose a delay-aware bandwidth slicing mechanism in PON-based mobile backhaul networks. The method tries to guarantee the transmission of migration traffic within the deadline, while at the same time minimizing the negative impact on non-migration traffic. Simulation results show that migration data can be transmitted successfully in a required time threshold, while the requirements of latency and jitter for non-migration traffic with different priorities can be well satisfied.

Acknowledgements

This work is supported in part by SJTU State Key Laboratory of Advanced Optical Communication System and Networks Open Project 2018GZKF03001, Swedish Research Council (VR) project 2016-04489 "Go-iData," Swedish Foundation for Strategic Research (SSF), and Chalmers ICT-seed grant.

Author details

Jun Li[1], Xiaoman Shen[2], Lei Chen[3*] and Jiajia Chen[1*]

1 Chalmers University of Technology, Göteborg, Sweden

2 Zhejiang University, Hangzhou, China

3 RISE Research Institutes of Sweden, Göteborg, Sweden

*Address all correspondence to: lei.chen@rise.se and jiajiac@chalmers.se

References

[1] 3GPP TS 122261, Service requirements for next generation new services and markets; V 15.5.0, 2019

[2] Chiang M, Zhang T. Fog and IoT: An overview of research opportunities. IEEE Internet of Things Journal. 2016; **3**(6):854-864

[3] Ku Y. 5G radio access network design with the fog paradigm: Confluence of communications and computing. IEEE Communications Magazine. 2017;**55**(4): 46-52

[4] Wang S, Xu J, Zhang N, Liu Y. A survey on service migration in Mobile edge computing. IEEE Access. 2018;**6**: 23511-23528

[5] Li J, Shen X, Chen L, Pham D, Ou J, Wosinska L, et al. Service migration in fog computing enabled cellular networks to support real-time vehicular communications. IEEE Access. 2019;**7**: 13704-13714

[6] Machen A, Wang S, Leung KK, Ko BJ, Salonidis T. Live Service Migration in Mobile Edge Clouds. IEEE Wireless Communications. 2018;**25**(1): 140-147

[7] Codeca L, Frank R, Engel T. Luxembourg SUMO Traffic (LuST) Scenario: 24 Hours of Mobility for Vehicular Networking Research. Kyoto: IEEE Vehicular Networking Conference (VNC); 2015. pp. 1-8

[8] Han D, Shin S, Cho H, Chung J, Ok D, Hwang I. Measurement and stochastic modeling of handover delay and interruption time of smartphone real-time applications on LTE networks. IEEE Communications Magazine. 2015; **53**(3):173-181

[9] Li J, Natalino C, Van D.V, Wosinska L, Chen J, Resource Management in Fog Enhanced Radio Access Network to Support Real-Time Vehicular Services. Madrid: IEEE Information Conference on Fog and Edge computing; May 2017

[10] Li J. Ultra-Low Latency Communication for 5G Transport Networks. Universitetsservice US AB; 2019. ISBN: 978–91–7873-243-2

[11] Morabito R, Cozzolino V, Ding AY, Beijar N, Ott J. Consolidate IoT edge computing with lightweight virtualization. IEEE Network. 2018; **32**(1):102-111

[12] Zhang H, Chen K, Bai W, Han D, Tian C, Wang H, et al. Guaranteeing deadlines for inter-data Center transfers. IEEE/ACM Transactions on Networking. 2017;**25**(1):579-595

[13] Kramer G. Ethernet Passive Optical Networks. New York: McGraw-Hill; 2005

[14] Dixit A, Lannoo B, Colle D, Pickavet M, Demeester P. Energy Efficient DBA Algorithms for TWDM-PONs. In: IEEE International Conference on Transparent Optical Networks (ICTON). Budapest: IEEE; 2015. pp. 1-5

[15] Kramer G, Mukherjee B, Pesavento G. IPACT: A dynamic protocol for an Ethernet PON (EPON). IEEE Communications Magazine. Feb., 2002;**40**(2):74-80

[16] Li J, Shen X, Chen L, Ou J, Wosinska L, Chen J. Delay-aware bandwidth slicing for service migration in mobile backhaul networks. IEEE/OSA Journal of Optical Communications and Networking. 2019;**11**(4):B1-B9

Chapter 3

Artificial Intelligence and Machine Learning in 5G and beyond: A Survey and Perspectives

Abdelfatteh Haidine, Fatima Zahra Salmam, Abdelhak Aqqal and Aziz Dahbi

Abstract

The deployment of 4G/LTE (Long Term Evolution) mobile network has solved the major challenge of high capacities, to build real broadband mobile Internet. This was possible mainly through very strong physical layer and flexible network architecture. However, the bandwidth hungry services have been developed in unprecedented way, such as virtual reality (VR), augmented reality (AR), etc. Furthermore, mobile networks are facing other new services with extremely demand of higher reliability and almost zero-latency performance, like vehicle communications or Internet-of-Vehicles (IoV). Using new radio interface based on massive MIMO, 5G has overcame some of these challenges. In addition, the adoption of software defend networks (SDN) and network function virtualization (NFV) has added a higher degree of flexibility allowing the operators to support very demanding services from different vertical markets. However, network operators are forced to consider a higher level of intelligence in their networks, in order to deeply and accurately learn the operating environment and users behaviors and needs. It is also important to forecast their evolution to build a pro-actively and efficiently (self-) updatable network. In this chapter, we describe the role of artificial intelligence and machine learning in 5G and beyond, to build cost-effective and adaptable performing next generation mobile network. Some practical use cases of AI/ML in network life cycle are discussed.

Keywords: Next Generation mobile Networks, 5G, Artificial Intelligence, Machine Learning, Deep Learning, Physical Layer, Big Data, Network Control

1. Introduction

The massive deployment of LTE (Long Term Evolution) or 4G mobile network has solved one of the major challenges of wireless communications, which is high capacities, to build real broadband mobile Internet. This was possible mainly through very strong physical layer, based on orthogonal frequency division multiplexing (OFDM) and multiple input multiple output (MIMO) among others, and flexible network architecture. However, new bandwidth-hungry services have been developed in unprecedented way, reaching capacities up to 1 Gbps, such as virtual reality (VR), augmented reality (AR), etc. Furthermore, mobile networks are facing other new services with extremely demand of higher reliability and almost zero-latency performance, like vehicle communications or Internet-of-Vehicles (IoV).

The 5G systems solved the major problems related to the capacity through use of new radio interface, massive MIMO, beamforming, high modulation orders, etc. Furthermore, 5G is planned to include a high level of flexibility to optimize the network utilization by integrating software defined networking (SDN) and network function virtualization (NFV) technologies. This should allow the network operators to support current and new more demanding future services. The main challenge is to be ready to support services for customers in completely different vertical markets/industries, like e-health, Internet-of-Vehicles (IoV), Industry 4.0, smart grids, etc. Furthermore, the network operators have to establish more partnerships on multiple layers for sharing of the 5G infrastructure through network sharing relationship among different mobile operators, delivery of Infrastructure as a Service, Platform as a Service or Network as a Service by assets providers. Facilitating such partnerships may act as catalyst for the deployment of 5G networks, considering the large investments for mobile network operators (MNOs) in capital expenditure CAPEX and operational expenditure OPEX are still not being followed by significant revenue increase, [1]. Network operators are also forced to consider a higher level of intelligence in their networks, in order to deeply and accurately learn the operating environment and users behaviors and needs. The adoption of Artificial Intelligence (AI), and Machine Learning (ML) approaches as core part of AI, is crucial to forecast the evolution of the environment and users/services behavior/demand to build a pro-actively and efficiently (self-) optimizing and (self-) updating networks. This is true for each layer of the system and each level of the network. For example, AI/ML are crucial for massive MIMO to identify dynamic change and forecast the user distribution by analyzing historical data, dynamically optimize the weights of antenna elements using the historical data or to improve the coverage in a multi-cell scenario considering the inter-site interference between multiple 5G massive MIMO cell sites, etc.

In this chapter, we describe the role and the integration method of AI and different ML approaches as core part of AI in the next generation mobile networks. The rest of the chapter is built by giving a short overview on AI and ML definitions, historic and (sub-) classes in second section. The third section shows the role of big data as prerequisite for a full exploitation of AI/ML advantages. The components of 5G are described in fourth section and how to make AI/ML a main component of next generation mobile networks. Practical use cases are illustrated and discussed in fifth section.

2. AI and ML in mobile communications networks

2.1 Overview on AI and ML

2.1.1 What is AI?

AI is the scientific field that deals with programming machines to mimic human behavior in solving tasks that humans are good at (natural language, speech, image recognition, etc.). AI involves the intersection of many fields of computer science and applied mathematics. The position of artificial intelligence is rather to consider that we, as human beings, have an intuitive understanding of what intelligence is and therefore we can judge whether a machine is intelligent or not. This operational definition of AI was promoted by Alan Turing in 1950, who introduced his famous "Turing test". The Turing test is an operational test; according to which a machine is considered intelligent if it can converse in such a way that (human) interrogators cannot distinguish it from a human being [2].

Initial efforts at AI involved modeling the biological neurons in the brain. In 1943 McCulloch and Pitts [3] modeled for the first time the artificial neural as a binary variable that is switched to either on or off. Later in 1949, Donald Hebb developed an algorithm for learning neural networks. In 1951, Marvin Minsky and Dean Edmonds built the Stochastic Neural Analog Reinforcement Calculator (SNARC), the first neural network computer. Following this accomplishment, a small group of scientists interested in the study of intelligence met in a 2-month workshop at Dartmouth University in 1956. According to common belief, the term AI was first introduced and defined by John McCarthy at this workshop, as:" AI involves machines that can perform tasks that are characteristic of human intelligence".

Over the last few decades, AI has gained increasing interest among researchers and industry. This is due to the wide variety of applications in which AI has been used, such for example, natural language processing (e.g. broadcast news transcription, speech-to-speech translation), healthcare (e.g. assisting in surgeries, computer aided diagnosis), smart cars and drones (e.g. self-driving cars, obstacle detection) and also mobile networks (e.g. performance optimization, traffic prediction).

2.1.2 The connection between machine learning, deep learning and AI

Today AI is a collection of different technologies working together to enable machines to sense, comprehend, act, and learn with human-like levels of intelligence. Rule-based techniques as well as expert system are the first approaches to AI. Technologies such as ML, Deep Learning, and Big Data are all part of the AI landscape, as illustrated in **Figure 1**. As the most recent advances in AI have been in the field of machine learning, people often mistakenly conflate AI with ML. Eventually, the field of MML was created from the desire to design AI with the ability to learn and acquire knowledge.

ML is a subset of AI, which aims to give the ability for a computer to perform tasks without being given explicit instructions on how to solve it. It is a paradigm that aims to build a computer that can learn, just like humans do. The learning

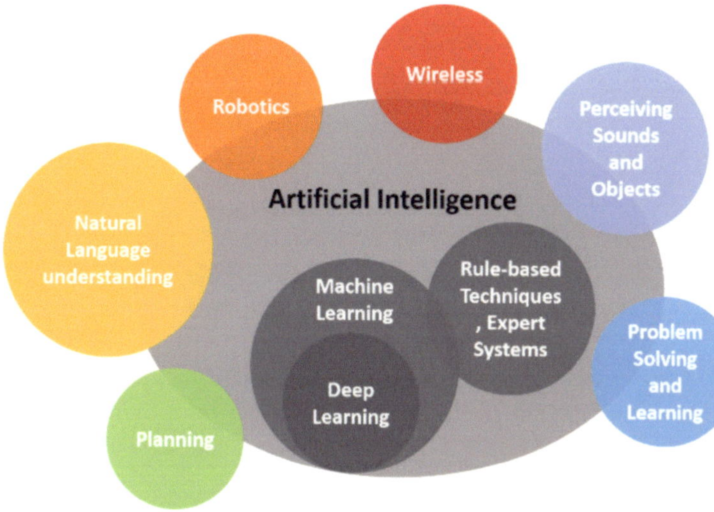

Figure 1.
Connection and overlap between machine learning, deep learning, and artificial intelligence.

process consist of providing a ML algorithm with examples of the task we want to solve (data), and letting the computer finding patterns and making inferences that optimizes the decision making according to a user-defined objective. In general, ML could be used to accomplish different type of tasks including classification, clustering and making predictions about data.

Artificial Neural Networks (ANN), referred also as Neural Networks (NN), are a popular machine learning models inspired by the biological processes of the brain. The first NN algorithm is the Perceptron developed by Rosenblatt in 1958; [4]. This finding was inspired by McCulloch mathematical models of neurons in the human brain [3]. In the following decades, different types and architectures of neural networks have been proposed as well as algorithms to train them effectively. Following these accomplishments, the term Deep Learning was introduced to the ML community. Around the year 2000, Deep Learning is a subcategory of ML focused on parameterizing multilayer (deep) neural networks that can learn representations of the data.

In recent years, deep learning based methods has gained increasing interest. This is due to their high performances in image classification [5], speech recognition [6] and natural language processing tasks [7]. In fact, deep learning techniques have largely outperformed stat-of-the-arts results in these tasks. However, deep learning, and indeed most of ML techniques, have several limitations. The first limitation is the amount of data required during training in order to achieve human like performances. Another limitation is the computational capacities required to train deep learning based models on large datasets.

2.1.3 The categories of machine learning

As pointed out in the previous paragraphs, ML is a complex landscape. Based on the training strategy, ML can be divided into three classical categories; which different learning approaches are illustrated in **Figure 2**:

Supervised Learning: Learning with a labeled training set. This category includes Classification and Regression tasks.

Unsupervised Learning: The process of training a model using training data that is unlabeled. The model had to discover patterns in unlabeled data. The widely used task in unsupervised learning is Clustering.

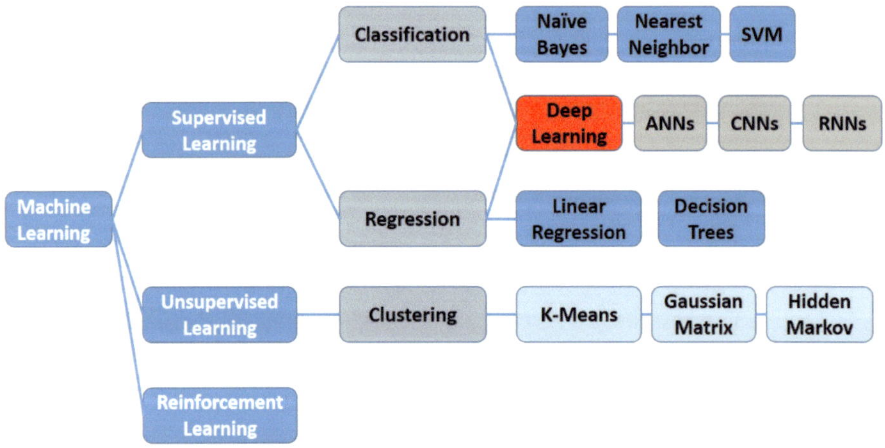

Figure 2.
Classification of different ML approaches.

Reinforcement Learning: The process of training a model on a series of actions that lead to a particular outcome, where the system receives rewards for performing well and punishments for performing poorly directly from its environment. Reinforcement Learning is used in robotics and games.

2.2 Introducing AI and ML in mobile communications

2.2.1 Needs for intelligence in mobile networks prior to 5G

As one of the worldwide leading mobile systems manufacturers, Ericsson has led a study to analyze the state-of-the-art and the expectation of adopting AI by the mobile network operators and global service providers. The study found that operators have mainly adopted AI as means to enable them to switch to 5G and to guarantee optimized investment; [8]. Furthermore, already with 4G/4.5G there has been an increased complexity in the management of a vast number of devices and huge amount of data. Operators are hoping that AI and ML will help to reduce this complexity. Other main findings of the study are as follows; [8]:

- AI is already being incorporated into networks, with a primary focus on reducing capital expenditure, optimizing network performance and building new revenue streams. Operators from all over the world are already reaping the benefits of integrating AI into their networks. More than half of service providers (53 percent) expect to have fully integrated some aspects of AI into their networks by the end of 2020.

- AI will be vital for improving customer service and enhancing customer experience, generally referred as "Quality of Experience (QoE)" to. AI is expected to help providers further improve customer experience in many ways, including improving network quality and providing personalized services.

- AI will help recoup the investments communications service providers (CSPs) are making in their networks to switch to 5G. Lowering operational costs and ensuring returns on network investments are key priorities that service providers are looking to achieve using Artificial Intelligence. **Figure 3** shows the prioritized domains for the integration of AI to optimize the costs as well as the management of the always-increasing network complexity. Network intelligence and automation are crucial to the evolution of 5G, IoT and industrial digitalization. As 5G-enabled technologies develop; operators will need to increase their network capacity. However, the added capacity brings additional complexity.

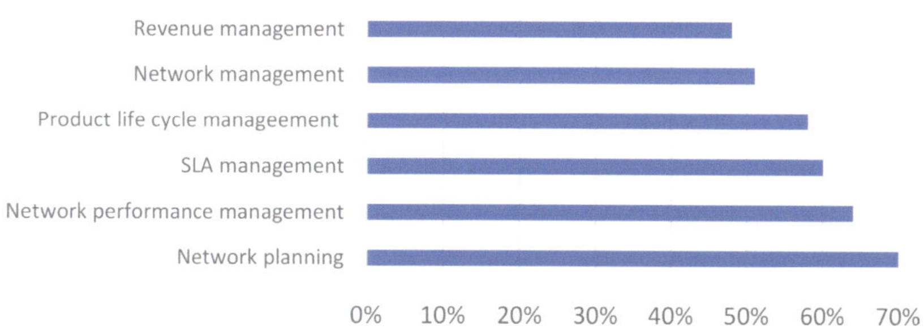

Figure 3.
Core areas with the highest potential for returns on AI investments [8].

- Adopting AI is creating new data challenges, even as it solves network complexities. Network providers agree that they need to develop effective mechanisms for collecting, structuring and analyzing the huge volumes of data that AI is capable of amassing.

2.2.2 Advantages of using AI in mobile networks

The history of mobile communication is evolving from one generation to a next one over the last three decades. Nowadays, before to jump to a next generation, i.e. 5G, the mobile network operator need to understand deeply the current situations and future scenarios. The MNO needs to learn more about the real behavior of the subscribers, their profiles, the traffic patterns, wished and possibly adequate services for each profile and how all these will evolve in the future. In the communications market, whether it is network operators, equipment manufacturers or solution providers, etc., the market players hope to take advantage of AI to assist in areas that become very challenging, such as in designing, operating, maintaining and managing communication networks and services. Mainly AI will offer the MNO the ability of learning more about their network and the need of customers, ability of understanding and reasoning to make the ideal decision/actions for different scenarios and environment conditions and finally the ability to collaborate between high heterogeneous widely expanded and densified network infrastructures.

Operators need intelligent decisions to manage complex resources and dynamic traffic. However, so far no one single model has the ability to model accurately the network traffic characteristics. Fortunately, AI has entered into the cognitive age, and deep learning can be used. Through deep learning, the machine system can use the existing training data to process large amounts of data through data mining. AI can also learn the characteristics of data traffic, management, controls and other characteristics automatically and master expert experience of operating, managing and maintaining networks. By these efforts, the accuracy of analysis can be enhanced, and the intelligent management and services of communication networks can be realized. Detailed description can be found in [9].

Due to the high dynamics of the network system, the state information of a resource may have changed when it is transmitted to the network management system. Therefore, the network management can only know the local state information without the knowledge about the system internal state. ML has the strength to deal with this kind of fuzzy logic and uncertainty reasoning. In order to make the classification or prediction of the state easier, deep learning constructs a multi-hidden layer model and uses the hierarchical network structure to transform the feature representation of the sample into a new feature space layer by layer, as detailed in the first sections of the chapter. The major advantage is the fact that AI does not need to describe the mathematical model of the system accurately, and therefore has the ability to deal with uncertainty or even 'unknowability.

Due to the expansion and high densification of the network infrastructure, both in scale and size, the structure complexity of communication networks, and especially for the next generation, are increasing quickly. Concepts such as distribution and hierarchy are often talked about in the network management. Management tasks and controls are distributed to the entire network, in order to avoid centralization of the management functions that requires more data overhead and reliability issues. As a result, network operators have to deal with issues such as tasks' distribution, communication and collaboration between management nodes. The introduction of the multi-agent collaboration of distributed AI into the network management will support the ability to collaborate between network managers distributed in every layer. However, such collaboration requires a high

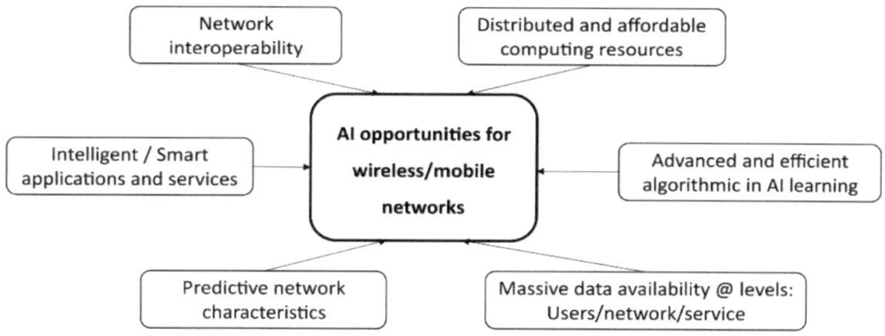

Figure 4.
Factors contributing to a full and efficient AI integration into mobile networks.

level of interoperability between the heterogeneous networks building the entire communications infrastructure. Furthermore, this network interoperability is just of the many other factors that must be fulfilled by the environment, in order to take benefits of the above-cited advantages of AI. The other factors allowing taking full potential from AI are depicted in **Figure 4**.

One of the major enablers of a profitable AI integration is the availability of high performing and efficient computing resources. Indeed, 5G systems seek to provide high throughput and ultra-low latency communication services, to improve users QoE. Implementing deep learning to build intelligence into 5G systems, to meet these objectives is expensive. This is because powerful hardware and software is required to support training and inference in complex settings. Several tools are emerging, which make deep learning in mobile networks tangible as discussed in [10]. Authors discussed a hierarchy of advanced computing tools as deep learning integration enablers; namely: (i) advanced parallel computing, (ii) distributed ML systems, (iii) dedicated deep learning libraries, (iv) fast optimization algorithms, and (v) fog computing.

3. Big data as prerequisite for integrating AI in mobile networks

The integration of intelligent algorithms and learning approaches requires the availability of big data sets, which represent the starting point. However, because the AI and ML can be integrated at different levels of the next generation mobile work, or at least in 5G, different big data sources can be explored and must be selected carefully to be able to extract as much as possible of knowledge/learning. Different ways are proposed in the literature to classify the needed types of big data sets, which should build the basis for the AI and ML, such the ones proposed in [10–12]. For example, a classification of different big data sources is illustrated in **Figure 5**, which contains three pools of big data sets: general wireless data, social network-aware data, social data and cloud data [11].

Wireless data: This class represents the big data generated by wireless users, and which contains useful information about the activity patterns in time, frequency and space (locations). For example, this can help to infer from the data traffic/demand variation over time, the interference power at different frequencies, and the congestion level distribution at different locations, etc. The exploitation of these spectral patterns will allow an efficient management of the wireless resources, at radio resource management/allocation (RRM/RRA) functionalities, for improving the systems spectral efficiency and for enhancing the delivered quality of service experienced by the end-user. One of the possible intelligent applications can be the load balancing relying on proactive RRA. In such context, the network operator

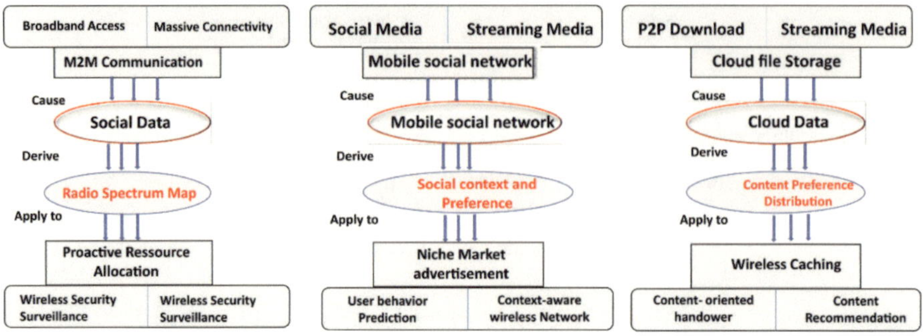

Figure 5.
Classes of big data in mobile network and their applications [11].

can adjust the transmit power, frequency or direction, through the beamforming mechanism adopted for 5G or simply through sectorized antennas, of the different base station transmitters relaying on the mobile users distributions. In addition, the operator can dispatch mobile base stations in advance when anticipating a regional surge of data traffic, which could be caused for example by sport events, music concerts, etc. In [10], authors subdivided the mobile big data into classes: Network level-data, which is similar to "wireless data", and App-level data. It is worth to notice that network level-data are further subdivided into sub-classes, namely:

- Infrastructure data: Infrastructure locations, capability, equipment holders, etc.
- Key performance indicator (KPI) data: Data traffic, data throughput, end-to-end delay, QoE, jitter, bit/packet error rates, etc.
- Call details records (CDR) data: Session start and end times (i.e. inter-arrival time, holding time), type, sender and receiver, etc.
- Radio information data: Signal power, frequency, spectrum, modulation, serving BS, etc.

Social data: Nowadays, the main cause of the soaring data volume in the Internet is the online social networks. The penetration of the mobile Internet into our daily lives and behaviors makes convenient multimedia communications easily accessible for everyone, independently of the age or social level. The volume of social data, which represents the data exchanged on social media, has reached an unprecedented astonishing magnitude and it is expected to continue an explosive increase in the next years. On one hand, its strong ties to public events in the physical world feature social network data. So that, an important football game or political event may inspire heated online discussions that may last for day in case of games and some weeks or months in case of an elections year. On the other hand, social network data contains rich information about the context/preferences of individuals or social groups. The mobile network operator can exploit such data to build a social-network-aware wireless concept. For example, operators may offer more bandwidth in touristic sites so that users/tourists can share their tourist-experience on the social media with highly satisfying quality of service even if they have only paid pre-paid service. Example of connecting the social media big data source to a AI learning framework is illustrated in a simple way in **Figure 6**, showing the most promising expected AI applications, like mobile caching, drone-based mobile BS

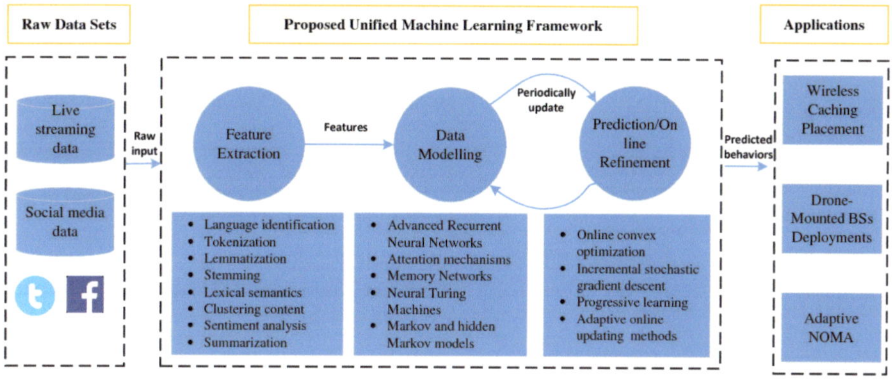

Figure 6.
Example of connecting big data source to learning framework before feeding the predicted behavior models to the AI applications layer [11].

placement or radio resource allocation in case of adaptive non-orthogonal multiple access (NOMA).

Cloud data: one of the major cause for the apparition of the data tsunami in mobile network is the transmission of the multimedia contents stored in the cloud servers. Furthermore, it was found that most contents transferred over Internet are based on its popularity. It is possible therefore to reduce the tele-traffic of the system by exploiting the users' preference for special cloud contents. For example, the most popular videos, for a specific geographic region, can pre-cached at the edge servers so that no real-time backhaul data downloading is needed for frequent request. In addition, individual user's preferences for specific multimedia contents can be used to predict the user's future content demand, based on which the network operator can perform a pre-feeding or recommendation actions.

4. AI and ML as main components for 5G

4.1 Main goals and components of 5G

While the evolution toward 4G/LTE was driven by more mobile data speed, the 5G system is confronted by more stricter and diverse requirements; as summarized in **Table 1**. The deployment of 5G systems seeks to provide high throughput and ultra-low communications latencies, to improve users' quality of experience (QoE). To meet these requirements, 5G targets three evolution axes to cope with the new applications fields; such autonomous cars/driving, industrial automation/smart manufacturing (Industrie 4.0), virtual reality, e-health, etc. These axes are:

Enhanced mobile broadband (eMBB): Allows for new bandwidth-hungry applications with extreme high data rate demands over a uniform coverage area. Examples include ultra-high-definition video streaming and virtual/augmented reality (VR/AR).

Massive machine-type communications (mMTC): A key characteristic of 5G communication services is the scalable connectivity demand for expanding the number of wireless devices with efficient transmission of small amounts of data over extended coverage areas. Applications like body-area networks, smart homes, IoT, and drone delivery will generate this type of traffic. mMTC must be able to support massive new uses and others uses that would appear in the future.

Ultra-reliable low-latency communications (URLLC): Connected healthcare, remote surgery, mission-critical applications, autonomous driving, unmanned aerial vehicles (UAV), vehicle-to-vehicle (V2V) communications, high-speed train

Requirement	Desired value	Application example
Data rate	1 to10Gbps	Virtual reality office
Data volume	9GB/h (busy period) 500GB/month/user	Stadium, Dense urban, information society
Latency	Less than 5 ms	Traffic efficiency and safety
Battery Life	One decade	Massive deployment of sensors and actuators
Connected devices	300,000 devices/AP	Massive deployment of sensors (massive IoT)
Reliability	99.999%	Teleprotection in smart grid network, Traffic efficiency and safety

Table 1.
Summary of major requirements for 5G mobile system [13].

connectivity, and smart industry applications. These applications prioritize high reliability, low latency, and mobility over data rates.

To achieve the above listed objectives, 5G systems use a very large number different advanced technologies or enablers; such as new PHY layer, new MAC layer, SDN, NFV, etc. To have a structured view of these different enablers, **Figure 7** classifies them into two main categories; the system-level and network-level technologies. More details about of the listed technologies and paradigms can be found in the survey [14]. If we consider the breakthrough of the unprecedented achieved capacities, three enabling technologies can be cited, namely: massive MIMO that allows a new level of bit rates, use of millimeter-wave that finally give a hope to overcome the spectrum scarcity and network densification.

The introduction of MIMO in 4G/LTE was decisive in achieving real broadband speeds for mobile Internet. With the same philosophy, the massive MIMO is bringing the next mobile network to the next level of throughputs up to Gbps. Therefore, massive MIMO is considered as a leading candidate technology for 5G, where the high number of antennas at the base station requires large number of power amplifiers. According to [15], the primary problem with power amplifiers is known as the trade-off between linearity and efficiency: amplifiers can be designed to attain good linearity at the cost of efficiency. As the highly linear power amplifiers are

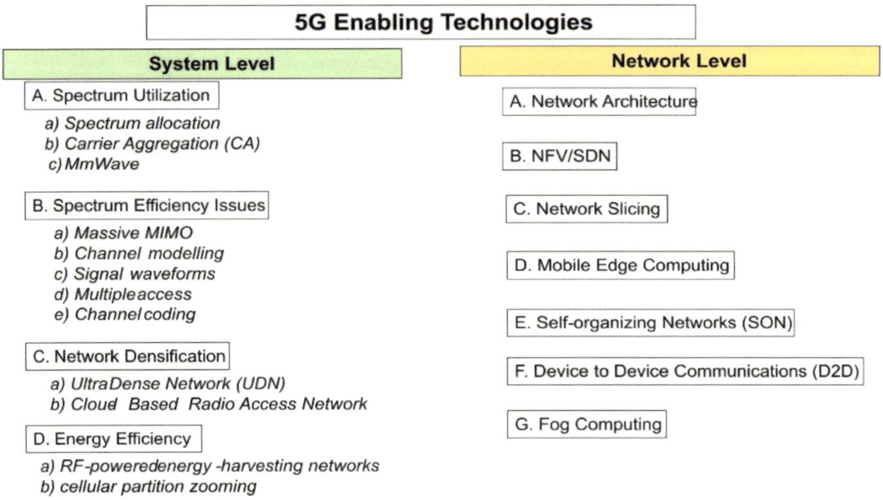

Figure 7.
Classification of major enabling technologies of 5G systems and networks.

expensive and power inefficient, the excessive use number of antennas at the base station makes the use of inexpensive elements strongly targeted to keep the overall network costs, capital expenditure and operational expenditure, manageable. However, emerging energy and spectrum-efficient wideband wireless communications systems are vulnerable to non-linear distortions that are attributed to the radio frequency front-ends. For example, those types of high-power amplifier affect the performance of the intended receiver and thus the entire network.

Furthermore, because the 5G systems should operate in more dynamic environments and in different bands (extending from cm-wave to mm-waves), the dynamic range requirements will likely become more demanding. Therefore, the power amplifiers need to meet stricter linearity specifications while at the same time maintaining an acceptable efficiency for the overall systems. The aimed high efficiency of power amplifiers is achievable when constantly feeding the amplifier at the limit of its high-power linear zone. However, this is not realistic for 5G base station, because this is not feasible solution for the peak-to-average power ratio.

The previously cited 5G enablers have a direct contribution in the network performance; however, an operative and efficient 5G network cannot be complete without AI. For example, 5G enables simultaneous connections to multiple IoT devices, generating massive amounts of data that must be processed using ML and AI. When ML and AI are integrated within, wireless providers can, for example [16]:

- Identify dynamic change and forecast the user distribution by analyzing historical data,
- Forecast the peak traffic, resource utilization and application types; and optimize and fine tune network parameters for capacity expansion,
- Eliminate coverage holes by measuring the interference and using the inter-site distance information,
- High level of automation from the distributed ML and AI architecture at the network edge,
- Application-based traffic steering and aggregation across heterogeneous access networks,
- Dynamic network slicing to address varied use cases with different QoS requirements,
- ML/AI-as-a-service offering for end users, etc.

4.2 Machine learning/deep learning in 5G

With the increasing advances and advantages of ML in the wireless communications, each research community has tried to evaluate the impact of ML on 5G in its discipline. As result, we have several publication for ML impact on physical layer, security aspects, radio resource managements, etc. This makes it very difficult to give short overview on the utilization and the impact of AI/ML in 5G. Therefore, we can summarize the works applying ML to 5G according to two principles: a "general ML categorization", where we consider all possible ML approaches from the literature, and a "Deep Learning-based Categorization", which focuses only the deep learning, because several leading publications consider deep learning as the most promising approach of ML for the high complexity of 5G.

Learning classes	Learning models	Example of applications in 5G
Supervised learning	ML and statistical logistic regression techniques.	Dynamic frequency and bandwidth allocation in self-organized LTE dense small cell deployments
	Support Vector Machines (SVM)	Path loss prediction model for urban environments
	Neural-Network-based approximation	Channel Learning to infer unobservable channel state information (CSI) from an observable channel
	Supervised ML Frameworks	Adjustment of the TDD Uplink-Downlink configuration in XG-PON-LTE Systems to maximize the network performance based on the ongoing traffic conditions in the hybrid optical-wireless network
	Artificial Neural Networks (ANN), and Multi-Layer Perceptrons (MLPs).	Modeling and approximations of objective functions for link budget and propagation loss for next-generation wireless networks
Unsupervised Learning	K-means clustering, Gaussian Mixture Model (GMM), and Expectation Maximization (EM).	Cooperative spectrum sensing and Relay node selection in vehicular networks.
	Hierarchical Clustering.	Anomaly/Fault/Intrusion detection in mobile wireless networks
	Unsupervised Soft-Clustering ML Framework.	Latency reduction by clustering fog nodes to automatically decide which low power node (LPN) is upgraded to a high power node (HPN) in heterogeneous cellular networks.
	Affinity Propagation Clustering.	Data-Driven Resource Management for Ultra-Dense Small Cells
Reinforcement Learning	Reinforcement Learning algorithm based on long short-term memory (RL-LSTM) cells.	Proactive resource allocation in LTE-U Networks, formulated as a non-cooperative game, which enables SBSs to learn which unlicensed channel, given the long-term WLAN activity in the channels and LTE-U traffic loads.
	Gradient follower (GF), the modified Roth-Erev (MRE), and the modified Bush and Mosteller (MBM).	Enable Femto-Cells (FCs) to autonomously and opportunistically sense the radio environment and tune their parameters in HetNets, to reduce intra/inter-tier interference.
	Reinforcement Learning with Network assisted feedback.	Heterogeneous Radio Access Technologies (RATs) selection.

Table 2.
Learning approaches and their 5G applications for the three ML classes.

A general ML categorization in case of 5G follows the general structure of ML as seen in the first sections, which uses three classes of ML: supervised learning, unsupervised learning and reinforcement learning. **Table 2** shows an example of such classification giving the used learning approaches from each class and a concrete example of application in 5G, [17]. In the next section, some 5G use cases will be described and solution for AI/ML integration in mobile network operators will be proposed.

Some research works focus only on the deep learning, because it is considered as most powerful learning approach of AI/ML. This gives a very large spectrum of applications in 5G and its different aspects, as detailed in the most recent surveys from examples [10, 18]. **Figure 8** shows just a small part of possible applications in

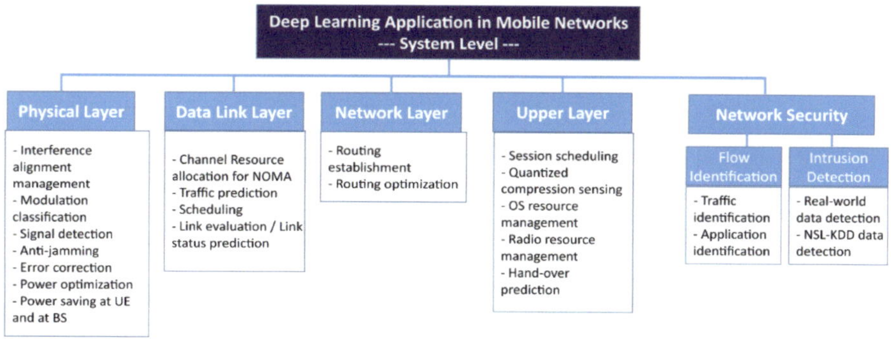

Figure 8.
Applications of deep learning in different layers of 5G systems (extended version of [18]).

5G systems and classify them according to each system layer where AI/ML is integrated. Furthermore, some works have focused on of the application of some very promising sub-variants of deep learning, like the application of Deep Q-learning for caching/offloading, network security and connectivity preservation, traffic engineering/resource scheduling, [19]. Deep learning presents several strengths to cope with the challenges in wireless communications, and especially in case of 5G, which are explained in detail in [10] and can be summarized as follows:

- Feature extraction: Deep neural networks can automatically extract high-level features through layers of different depths. This allows reducing the expensive hand-crafted feature engineering in processing heterogeneous and noisy mobile big data.

- Big data exploitation: Unlike traditional ML tools, the performance of deep learning usually grow significantly with the size of training data. Therefore, it can efficiently utilize huge amounts of mobile data generated at high rates.

- Unsupervised learning: Deep learning is effective in processing un−/semi-labeled data, enabling unsupervised learning. This is very important in handling large amounts of unlabeled data, which are common in mobile system.

- Multi-task learning: Features learned by neural networks through hidden layers can be applied to different tasks by transfer learning. This reduces computational and memory requirements when performing multi-task learning in mobile systems.

- Geometric mobile data learning: Dedicated deep learning architectures exist to model geometric mobile data, which revolutionize geometric mobile data analysis.

5. Use cases: network planning, optimisation and management

5.1 AI in network life cycle: from planning to control and management

The life cycle of a communication network starts with its planning, dimensioning and deployment in a first life phase. Here, the network operator aims an investment optimization, minimal capital expenditure (CAPEX), while respecting the design

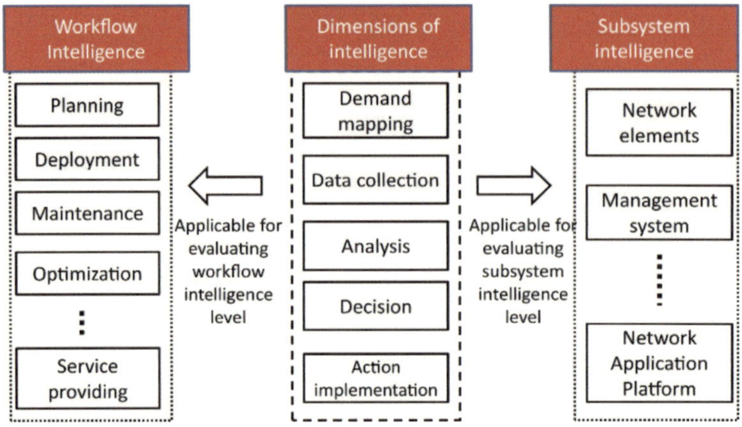

Figure 9.
Three pillars for integrating AI in network life cycle, by ITU-T [20].

requirements, especially the quality of service delivered or experimented by the end-users. In the second phase of the network life, a continuous control and management should guarantee a continuity of the service in a certain quality and a reliability of the services. In addition, network optimization must assist the management in order to keep the quality of service when necessary through upgrade the network hardware/software components to cope with the changes in the operating environment. Such changes can be in form of increase of the subscribers' number along the years or the apparition of new services with high demand of capacity, etc. In order to integrate a certain level of artificial intelligence in the above cited workflow processes (planning, dimensioning, etc.), **Figure 9** elaborated by the ITU-T illustrates the different intelligence pillars needed in the network intelligence landscape 19]. Therefore, the intelligence in the workflow requires different big data sets, like the demand mapping (and/or its forecasting for the coming years), a continuous collection of large data volume for the optimization tasks during the entire life cycle of the network. Moreover, the execution of such AI decisions and outputs requires intelligent sub-systems, which are able to interpret and to learn from the collected and analyzed big data streams.

The ideal case is to reach ZSM (Zero-touch network & service management), which is based on self-optimisation and self-healing network at different level of complexity either in the optimisation or in control and management. ITU-T gives three most desired cases, namely:

- Radio resource management for network slicing: Providing performance guarantee with high reliability, while ensuring efficient utilization of radio resources. For this purpose, the network should support the continuous collection of data, analysis of network slice behaviour and resource utilization patterns.

- End-to-end network service design automation: Automatically translating service requirements of application services to network parameters/requirements. Here also the big data sets are necessary. Therefore, network has to support data models to specify service requirements, integrate automated network configuration methods.

- End-to-end fault detection and recovery: Predictive detection and root cause analysis, and automated recovery decision making. This requires the collection of performance data on real-time basis, as well as generation of training data using testing environments.

Network Intelligence Level		Dimension of Intelligence				
		Action Implementation	Data Collection	Analysis	Decision	Demand Mapping
L0	Manual Operation	Human	Human	Human	Human	Human
L1	Assisted Operation	Human & System	Human & System	Human	Human	Human
L2	Preliminary Intelligence	System	Human & System	Human & System	Human	Human
L3	Intermediate Intelligence	System	System	Human & System	Human & System	Human
L4	Advanced Intelligence	System	System	System	System	Human & System
L5	Full Intelligence	System	System	System	System	System

Table 3.
Five possible degrees of intelligence in next generation networks [20].

However, it is clear that the migration toward a full intelligent network will not be an easy and one-dimensional task. Therefore, ITU-U has defined different degrees or levels of network intelligence according to five dimension, as listed in **Table 3** [20]. It is up to the network operator to define its own roadmap by prioritizing its objective concerning the investment, introduction of new services, etc. Nevertheless, in order to take a full benefit of the AI the operator should reach the fifth level "L5: Full Intelligence" in all the five dimensions. In addition, from telco's perspective the migration has to be in coherence with the business model as well as the return on investments.

5.2 AI/ML revolutionizing the planning and optimization process

One of the most critical processes to determine the final performance of a mobile network, and about its success in technical as well as financial aspects, is the initial planning process. Because the initial planning will determine also the way of functioning, operating, control and management processes, a bad-dimensioned network will always require more interventions from the control and management teams to try to bring the network performance to an acceptable level. In this process, decisions must be made about infrastructure (node deployment), spectrum, parameters and configuration setting procedures, energy consumption, network capacities to serve the worst-cases (peak or busy-hours traffic), evolution of the bandwidth demand over the years, etc. Furthermore, the planning and deployment of the next generation mobile network is not a greenfield task, i.e. starting from scratch. In fact, this planning task should take into consideration the already existing legacy systems and assets, such as point-of-presence, already existing base station, optical fiber for connecting the core network elements, the data center, etc.

In this very complex optimization problem, i.e. network planning, where several input parameters are uncertain random variables or distribution the AI can play a very important role in mastering the high complexity as well as delivering high efficient solutions. **Figure 10** shows how the AI can be integrated in the planning process [1]. The AI integrating module contains three parts:

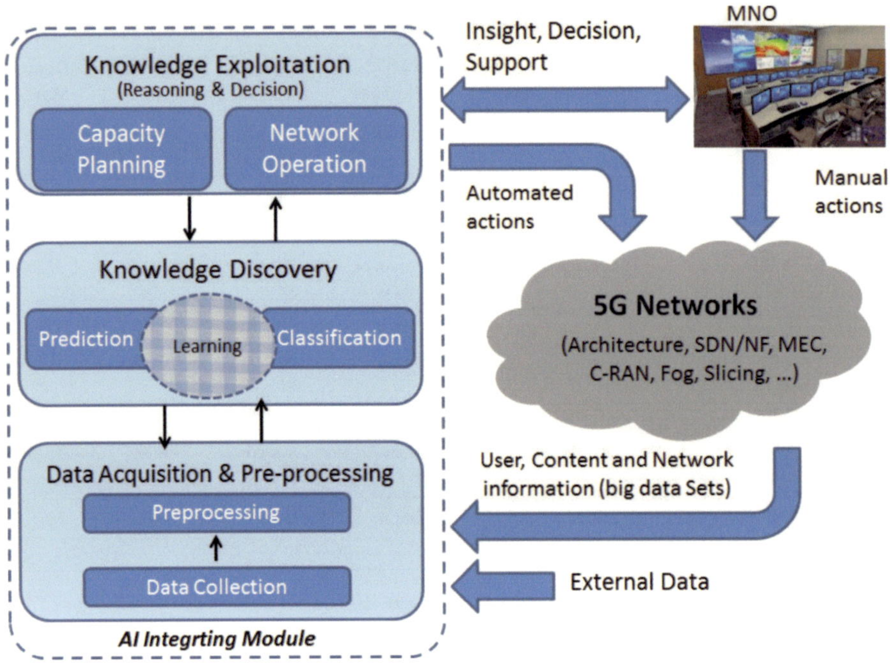

Figure 10.
Integrating AI/ML in the planning process of mobile networks (adapted from [1]).

- Data Acquisition and Pre-processing: Mobile Network Operators (MNOs) operate generally with complex, disparate sets of data, with useful information residing in multiple systems such as Operation and Management Systems (OMS), billing systems, inventories, network elements, Customer Relationship Management (CRM) systems, etc. However, to gain the challenge for achieving high performing future mobile network MNOs are forced to adopt efficient big data tools to bring together all necessary and profitable data sets. An AI-based planning system should be able to smartly analyze and correlate all these different data sources. A smart and efficient data management has to include all the necessary functions of collecting data, cleaning data, filtering data, correlating data from multiple sources and finding the relevant data.

- Knowledge discovery: this is the learning stage, where we try to learn and understand the traffic pattern and congestion, user behavior, resource usage, QoS, future location, faulty/problem elements/areas and their impacts on network efficiency. This stage should help to understand network performance and QoE, identify network anomalies, perform optimization, predict performance, disruption and requirements automate the control and operation, for example, self-configuration, self-optimization, self-healing.

- Knowledge exploitation: this stage will make use of the extracted knowledge (e.g. the prediction of the traffic patterns analysis in time/frequency/spectrum, the identified users' behaviors, etc.) to make decisions about the actions to be applied to the network elements/configurations. Such actions may either correct some system errors or improve the performance to be adapted to current/upcoming

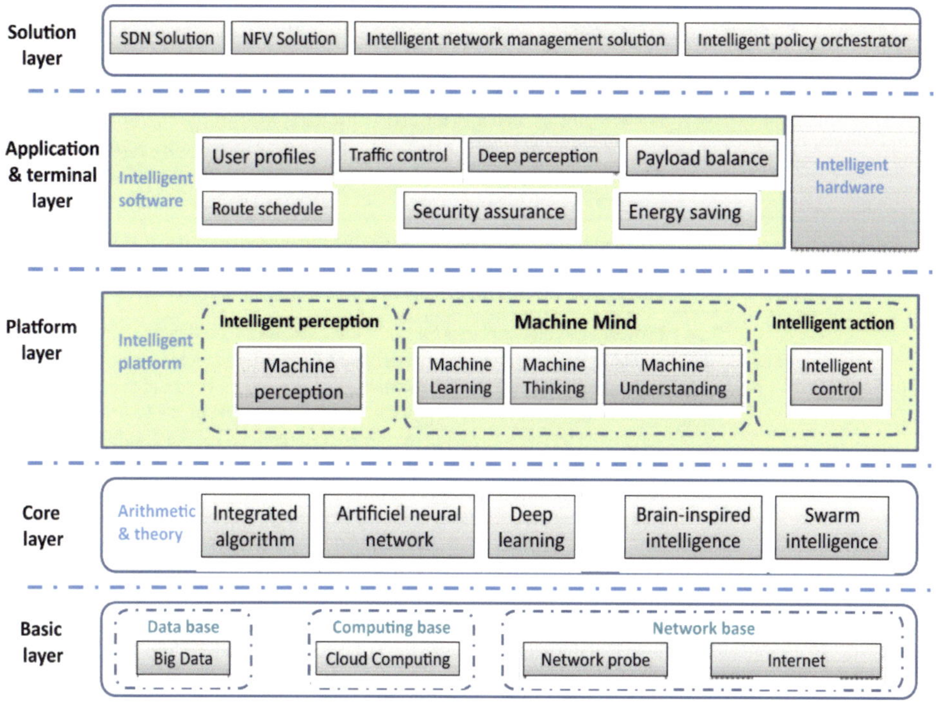

Figure 11.
Intelligence plane for the integration of AI in SDN, NFV and network control in the platform "FINE" [9].

situations and scenarios in the operating environment. In other words, this part gives out options and/or planning for slicing, virtualization, edge computing and impact of each decision option and/or planning, decisions about network expansion plan or resource utilization plan, suggestions on corrective actions.

5.3 AI building efficient collaboration NFV/SDN and network management

Already with 4G, the mobile network operators were facing an increasing network densification as response to the increasing demand for capacity and coverage, while with 4.5G operators were facing an exponential increasing number of end-devices, essentially in case of M2M and NB-IoT LTE. Therefore, research works have been dealing with the integration of AI in different levels of mobile architecture; independently of the access technology, either 4G or 5G. For example, authors in [9] proposed a functional architecture of the integration of AI to exploit and serve SDN, NFV and network control/monitoring. The authors proposed a framework of an intelligent communication network, called future intelligent network (FINE). The framework architecture is constituted of three planes: intelligence plane, agent plane and business lane.

In this section, we focus on the integration of the AI in SDN/NFV and network management, which is achieved through the intelligence plane that acts as the brain of the entire framework, **Figure 11**. Therefore, FINE is an intelligent network with an AI core. The intelligence plane can be composed of the basic layer, the core layer, the platform layer, the application/terminal layer and the solution layer. The basic tasks of each layers are summarized in **Table 4**.

Layers	Main tasks
Basic Layer	Provides support in data, calculation and the network for the intelligent plane. The data here is big data, not only including static data such as expert knowledge data, network infrastructure data, user profile data and others, but also including dynamic original data collected by the network probes from the business layer, such as status data of various types of equipment, applications and services.
Core Layer	Provider of intelligent algorithms in the intelligent plane, such as integrated algorithms, an artificial neural network, depth learning, brain-inspired intelligence and swarm intelligence. It is the kernel of the FINE core.
Platform Layer	Provides intelligent planes for the realization of the intelligent logic of AI ability and behaviour, such as intelligent perception, machine mind, intelligent action etc. The intelligent perception function can make use of theories and algorithms of the core layer, and deal with the big data of the basic layer supported by the computing resources, so as to perceive the development trends of networks and services. The machine mind function includes machine learning, machine thinking, machine understanding, etc. The ML consists of machine learning abilities generated by algorithms such as deep learning, brain-inspired intelligence and swarm intelligence. The machine thinking function provides the ability of knowledge mapping and knowledge reasoning. The machine understanding function provides the abilities of understanding based on the existing knowledge and the phenomenon, solving the ambiguity problem in reasoning, etc.
Application & Terminal Layer	Provides abilities of modular realization of functions needed by the solution layer. The functions here may include the user portrait, the flow control, the load balancing, the depth perception, the routing, the security, the energy saving, etc. These realizations may be in software or hardware using the abilities of perception, thinking and action provided by the platform layer.
Solution Layer	In charge of designing flexible policies and related activities related to satisfy the requirements to operate or manage the network, the network element, the network management system, etc.

Table 4.
Main tasks and layers in the intelligence plane of the platform "FINE" [9].

6. Conclusion

After building 4G network, network operators offered real broadband mobile Internet with capacities up to 600 Mbps. However, services and subscribers requirements have evolved to a very demanding and unprecedented level for higher quality of service. Virtual reality and augmented reality are demanding extreme high capacities and Internet of vehicles are requiring ultra-reliable communication and extreme low latency. This pushed the mobile network operators to start the migration toward 5G. Indeed, evolved technologies allowed 5G to reach bit rates over 1Gbps, through new radio interface, massive MIMO, beamforming, etc. However, the operators has to increase also the intelligence in their network, to learn more concisely about their operating environment and forecasting its evolution to optimize the resources utilization, adapt and configure automatically the network to cope with the wide variety of services. In this chapter, we have shown that this became possible through the integration of artificial intelligence and different machine learning approaches. We have presented some interesting use cases, which allow the operator to build self-healing and self-upgrading networks.

Nomenclature

4G/5G	Fourth/fifth generation of mobile network
AI	Artificial Intelligence
CAPEX	Capital Expenditure
Gbps	Giga bit per second.
LTE	Long Term Evolution
M2M	Machine-to-Machine communication
MAC	Medium Access Control
MIMO	Multi-Input Multi-Output
ML	Machine Learning
mMTC	Massive Machine-Type Communications
MNO	Mobile Network Operator
NB-IoT	Narrowband Internet of Things
NFV	Network Function Virtualization
NOMA	Non-Orthogonal Multiple Access
OFDM	Orthogonal frequency Division Multiplexing
OPEX	Operational Expenditure
QoE	Quality of Experience
QoS	Quality of Service
RRM/RRA	Radio Resource Management/Allocation
SDN	Software-Defined network

Author details

Abdelfatteh Haidine[1*], Fatima Zahra Salmam[2], Abdelhak Aqqal[1] and Aziz Dahbi[1]

1 National School of Applied Sciences, Chouaib Doukkali University, El Jadida, Morocco

2 Faculty of Sciences, Chouaib Doukkali University, El Jadida, Morocco

*Address all correspondence to: haidine.a@ucd.ac.ma

References

[1] J. Pérez-Romero, O. Sallent, R. Ferrús and R. Agustí, "Artificial Intelligence-based 5G network capacity planning and operation," 2015 International Symposium on Wireless Communication Systems (ISWCS), Brussels, 2015, pp. 246-250, doi: 10.1109/ISWCS.2015.7454338.

[2] Turing, A., Computing Machinery and Intelligence,"Mind, Vol. 59, 1950

[3] McCulloch, W., and W. Pitts, A Logical Calculus of the Ideas Immanent in Nervous Activity, Bulletin of Mathematical Biophysics, Vol. 5, 1943.

[4] Rosenblatt, F., The Perceptron: a Probabilistic Model for Information Storage and Organization in the Brain," Psychological Review,Vol. 65, No. 6, 1958

[5] Kaiming He, Xiangyu Zhang, Shaoqing Ren, and Jian Sun. Deep residual learning for image recognition. In CVPR, pages 770-778, 2016.

[6] Geoffrey Hinton, Li Deng, Dong Yu, George E Dahl, Abdel-Rahman Mohamed, Navdeep Jaitly, Andrew Senior, Vincent Vanhoucke, Patrick Nguyen, Tara N Sainath, et al. Deep neural networks for acoustic modeling in speech recognition: The shared views of four research groups. IEEE Signal Processing Magazine, 29(6):82-97, 2012.

[7] Ilya Sutskever, Oriol Vinyals, and Quoc V Le. Sequence to sequence learning with neural networks. In NIPS, pages 3104-3112, 2014.

[8] Ericsson AB: "Employing AI techniques to enhance returns on 5G network investments Ericsson AI & Automation Report, 2019. Online available https://www.ericsson.com/en/networks/offerings/network-services/ai-report (last retrieved 31.01.2021)

[9] G. Xu, Y. Mu and J. Liu: "Inclusion of Artificial Intelligence in Communication Networks and Services," ITU Journal: ICT Discoveries, Special Issue No. 1, 13 Oct. 2017

[10] C. Zhang, P. Patras and *H. Haddadi*, "Deep Learning in Mobile and Wireless Networking: A Survey," in IEEE Communications Surveys & Tutorials, vol. 21, no. 3, pp. 2224-2287, third quarter 2019, doi: 10.1109/COMST.2019.2904897.

[11] Y. Liu, S. Bi, Z. Shi and L. Hanzo, "When Machine Learning Meets Big Data: A Wireless Communication Perspective," in IEEE Vehicular Technology Magazine, vol. 15, no. 1, pp. 63-72, March 2020, doi: 10.1109/MVT.2019.2953857.

[12] M. G. Kibria, K. Nguyen, G. P. Villardi, O. Zhao, K. Ishizu and F. Kojima, "Big Data Analytics, Machine Learning, and Artificial Intelligence in Next-Generation Wireless Networks," in IEEE Access, vol. 6, pp. 32328-32338, 2018, doi: 10.1109/ACCESS.2018.2837692.

[13] FP7 ICT project: Mobile and Wireless Communications Enablers for the Twenty-Twenty Information Society 5G. [Online]. Available: https://www.metis2020.com/

[14] S. Hassani, A. Haidine and H. Jebbar: "Road to 5G: Key Enabling Technologies," Journal of Communications Vol. 14, No. 11, November 2019

[15] M. Yao, M. Souhoul. V. Marojevic and J. Reed: "Artificial Intelligence-Defined 5G Radio Access Networks", in IEEE Communications Magazine, vol. 57, no. 3, pp. 14-20, March 2019, doi: 10.1109/MCOM.2019.1800629.

[16] O. Dharmadhikari: "Leveraging Machine Learning and Artificial Intelligence for 5G", [Internet]. Avaiblae

from : https://www.cablelabs.com/leveraging-machine-learning-and-artificial-intelligence-for-5g (last accessed: 11/01/2021)

[17] M. E. M. Cayamcela and W. Lim, "Artificial Intelligence in 5G Technology: A Survey," 2018 International Conference on Information and Communication Technology Convergence (ICTC), Jeju, 2018, pp. 860-865, doi: 10.1109/ICTC.2018.8539642.

[18] Q. Mao, F. Hu and Q. Hao, "Deep Learning for Intelligent Wireless Networks: A Comprehensive Survey," in IEEE Communications Surveys & Tutorials, vol. 20, no. 4, pp. 2595-2621, Q4 2018, doi: 10.1109/COMST.2018.2846401.

[19] N. C. Luong et al., "Applications of Deep Reinforcement Learning in Communications and Networking: A Survey," in IEEE Communications Surveys & Tutorials, vol. 21, no. 4, pp. 3133-3174, 4th Q 2019, doi: 10.1109/COMST.2019.2916583.

[20] V. P. Kafle: "Network control with AI/ML – Standardization progress in ITU –," Presentation of National Institute of Information and Communications Technology (NICT), for the Institute of Electronics, Information and Communication Engineers (IEICE), Japan, 2020 May 21. Online available https://www.ieice.org/~nv/NV2020-01-Kafle.pdf. (Last retrieved 05.01.2021)

Chapter 4

A Brief Overview of CRC Implementation for 5G NR

Hao Wu

Abstract

In fifth generation (5G) new radio (NR), the medium access control (MAC) layer organizes the data into the transport block and transmits it to the physical layer. The transport block consists of up to million bits. When the transport block size exceeds a threshold, the transport block is divided into multiple equal size code blocks. The code block consists of up to 8448 bits. Both the transport block and the code block have a cyclic redundancy check (CRC) attached. Due to the difference in the size of the transport block and the code block, the CRC processing scheme suitable for the transport block and that suitable for the code block are different. This chapter gives an overview of the CRC implementation in 5G NR.

Keywords: 5G, NR, CRC, transport block, code block

1. Introduction

In order to provide high data transmission rates, the bandwidth of mobile communication systems is increasing. In fourth generation (4G) long term evolution (LTE), the maximum transmission bandwidth for one component carrier is 20 MHz [1]. In fifth generation (5G) new radio (NR), the frequency bands are divided into two parts: frequency range 1 (FR1) below 6 GHz and frequency range 2 (FR2) above 24.25 GHz. The maximum transmission bandwidth for one component carrier is 100 MHz and 400 MHz in FR1 and FR2 respectively [2]. The increasing system bandwidth brings new problems to the design of the transmitter and the receiver. In this chapter of the book, we focus on the cyclic redundancy check (CRC) implementation in 5G NR.

In 5G NR, there are many physical channels defined in the downlink and the uplink [3]. The downlink physical channels consist of the physical downlink shared channel (PDSCH), the physical downlink control channel (PDCCH), the physical broadcast channel (PBCH), etc. The uplink physical channels consist of physical uplink shared channel (PUSCH), the physical uplink control channel (PUCCH), the physical random access channel (PRACH), etc. The PDSCH and the PDSCH are mainly used to transmit data. The usage scenarios of 5G NR consist of enhanced mobile broadband (eMBB), massive machine-type communications (mMTC) and ultra-reliable and low latency communications (URLLC) [4, 5]. The usage scenario of the eMBB requires high data transmission rates. As a consequence, we focus on the PDSCH and the PUSCH in this chapter.

The medium access control (MAC) layer organizes the data into the transport block and transmits it to the physical layer. In 5G NR, the maximum transport block

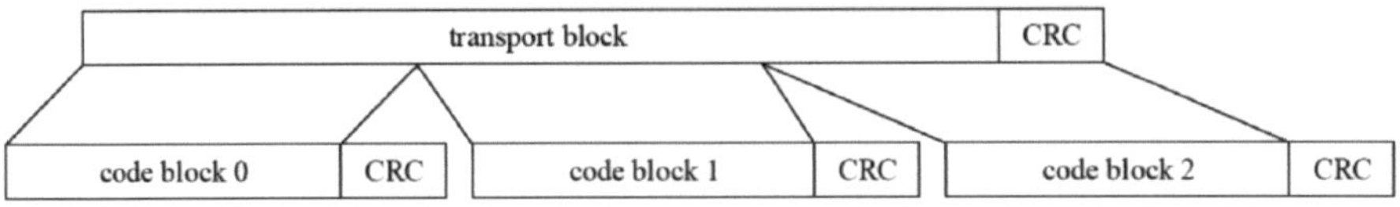

Figure 1.
The transport block and the code block.

size is 1,277,992 [6]. The processing of the transport block is shown in **Figure 1** [7]. If the transport block size is larger than 3824, a 16-bit CRC is added at the end of the transport block. Otherwise, a 24-bit CRC is added at the end of the transport block. The transport block is divided into multiple equal size code blocks when the transport block size exceeds a threshold. For quasi-cyclic low-density parity-check code (QC-LDPC) base graph 1, the threshold is equal to 8448. For QC-LDPC base graph 2, the threshold is equal to 3840. In 5G NR, the maximum code block size number is 8448. An additional 24-bit CRC is added at the end of each code block when there is a segmentation. Due to the difference in the size of the transport block and the code block, the CRC processing scheme suitable for the transport block and that suitable for the code block are different.

The rest of this chapter is organized as follows. Section 2 describes the system model of the transport block and the code block in 5G NR. Section 3 gives two properties of the CRC. Section 4 presents the overview of the CRC implementation. Finally, Section 5 gives the conclusion.

2. System model

Let $\boldsymbol{a} = [a_0, a_1, \dots, a_{L-1}, a_L, a_{L+1}, \dots, a_{L+N-1}]$ be the transport block including the transport block level CRC, where L is the transport block size and N is the transport block level CRC size. Note that $\boldsymbol{p} = [a_L, a_{L+1}, \dots, a_{L+N-1}]$ is the transport block level CRC. If L is smaller than or equal to 3824, then N is equal to 16 and $\boldsymbol{p}$ is generated by the following cyclic generator polynomial:

$$g_{16}(x) = x^{16} + x^{12} + x^5 + 1 \tag{1}$$

If L is larger than 3824, then N is equal to 24 and $\boldsymbol{p}$ is generated by the following cyclic generator polynomial:

$$g_{24\text{A}}(x) = x^{24} + x^{23} + x^{18} + x^{17} + x^{14} + x^{11} + x^{10} + x^7 + x^6 + x^5 + x^4 + x^3 + x + 1 \tag{2}$$

When $L + N$ is larger than M, the transport block including the transport block level CRC is segmented into multiple code blocks. Let R be code rate of the initial transmission indicated by the modulation and coding scheme (MCS) index. If $L > 292$ and $R > 0.67$ or $L > 3824$ and $R > 0.25$, then QC-LDPC base graph 1 is used and M is equal to 8448. Otherwise, QC-LDPC base graph 2 is used and M is equal to 3840.

When there is no segmentation, the number of code blocks C is equal to 1. When there is a segmentation, the number of code blocks C is equal to

$$C = \lceil (L + N)/(M - 24) \rceil \tag{3}$$

In the following sections, we mainly consider the case that there is a segmentation. Let $\boldsymbol{c}^i = \left[c_0^i, c_1^i, \dots, c_{K-1}^i\right]$ be the ith code block, where K is the code block size and is equal to

$$K = (L + N)/C + 24 \tag{4}$$

Note that the procedure of the transport block size determination guarantees that $(L + N)$ is divisible by C. $\left[c_{K-24}^i, c_{K-23}^i, \dots, c_{K-1}^i\right]$ is the code block level CRC, which is generated by the cyclic generator polynomial

$$g_{24\text{B}}(x) = x^{24} + x^{23} + x^6 + x^5 + x + 1 \tag{5}$$

c_j^i is equal to

$$c_j^i = a_{i(K-24)+j} \tag{6}$$

where $0 \leq j \leq K - 25$. In the following, the processing of the transport block includes: QC-LPDC encoding, rate matching, bit interleaving and code block concatenation. The encoded transport block is transmitted over the air after the symbol level processing.

At the receiver side, the following steps are carried out for the transport block: code block segmentation, bit de-interleaving, de-rate matching, QC-LPDC decoding, code block concatenation. We need to check whether each code block and the transport block are correctly received. Let $\boldsymbol{d}^i = \left[d_0^i, d_1^i, \dots, d_{K-1}^i\right]$ be the ith received code block after the hard decision and $\boldsymbol{e} = \left[e_0, e_1, \dots, e_{L+N-1}\right]$ be the received transport block after the hard decision. e_j is equal to

$$e_j = d_u^v \tag{7}$$

where $v = \lfloor j/(K - 24) \rfloor$, $u = \text{mod}(j, K - 24)$ and $0 \leq j \leq L + N - 1$. The undetected error probability is required to be less than 10^{-6} in 5G NR [8, 9]. Since the parity check capacity of QC-LDPC codes alone cannot meet the undetected error probability requirement of 5G NR [8, 9], we need to use the CRC check to determine whether $\boldsymbol{d}^i$ and $\boldsymbol{e}$ are correctly received.

3. Properties of the CRC

In this section, we give two properties of the CRC. These properties are useful in the CRC implementation. Before giving these properties, we define some variables. Let $A(x)$ and $B(x)$ be the polynomials. Let $g(x)$ be the cyclic generator polynomial. $\text{CRC}_{g(x)}[A(x)]$ is defined as the remainder when $A(x)$ is divided by $g(x)$. The two properties are listed as follows.

Property 1.

$$\text{CRC}_{g(x)}[A(x)B(x)] = \text{CRC}_{g(x)}\left[\text{CRC}_{g(x)}[A(x)]\text{CRC}_{g(x)}[B(x)]\right] \tag{8}$$

Property 1 implies that $\text{CRC}_{g(x)}[A(x)B(x)]$ can be obtained by computing the CRC of $A(x)$ and $B(x)$ independently.

Property 2.

$$\mathrm{CRC}_{g(x)}[A(x) + B(x)] = \mathrm{CRC}_{g(x)}[A(x)] + \mathrm{CRC}_{g(x)}[B(x)]. \tag{9}$$

Property 2 implies that $\mathrm{CRC}_{g(x)}[A(x) + B(x)]$ can be obtained by computing the CRC of $A(x)$ and $B(x)$ independently.

The proof of the property 1 and the property 2 can be found in Refs. [10, 11]. It is omitted for brevity. $g(x)$ in the expression of $\mathrm{CRC}_{g(x)}[A(x)]$ is clear from the context. As a consequence, $g(x)$ in the expression of $\mathrm{CRC}_{g(x)}[A(x)]$ is omitted in the following.

4. Overview of the CRC implementation

In this section, we give an overview of the CRC implementation. In the following, the received transport block after the hard decision $\boldsymbol{e}$ is used as an example. The implementation is easily generalized to other cases.

4.1 CRC implementation by direct calculation

In this scheme, the CRC of $\boldsymbol{e}$ is directly calculated by the division of polynomial using modulo-2 arithmetic.

Figure 2 illustrates an example. The dividend is equal to $x^5 + x^4 + x + 1$ and the divisor is equal to $x^2 + x + 1$. The division of polynomial begins by putting $x^5 + x^4 + x^3$ below $x^5 + x^4$. Subtracting and bringing down the next term give us the intermediate variable $x^3 + x$. This process is repeated until the degree of the intermediate variable is less than 2. Finally, we obtain that the quotient is equal to $x^3 + x + 1$ and the remainder is equal to x. That is,

$$\mathrm{CRC}\left[x^5 + x^4 + x + 1\right] = x \tag{10}$$

The division of polynomial using modulo-2 arithmetic is a computationally intensive operation. In the worst case, it requires a shift operation and an XOR logic operation for each bit of $\boldsymbol{e}$. As a consequence, this scheme is rarely used in actual systems. In order to solve the problem of the direct calculation, many schemes have been proposed in the literatures.

For example, the CRC implementation for $g(x) = x^5 + x^3 + x + 1$ is shown in **Figure 3** [12, 13]. The parallelism of this CRC implementation is 1 and thus one bit is

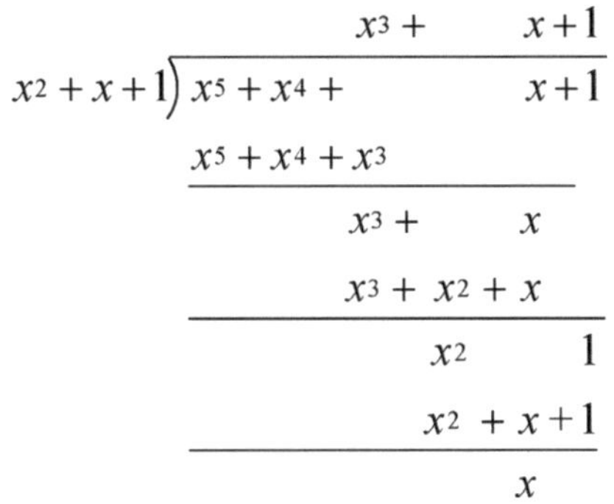

Figure 2.
The division of polynomial using modulo-2 arithmetic.

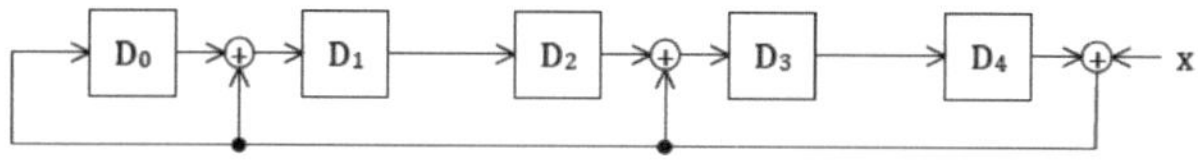

Figure 3.
CRC implementation for $g(x) = x^5 + x^3 + x + 1$.

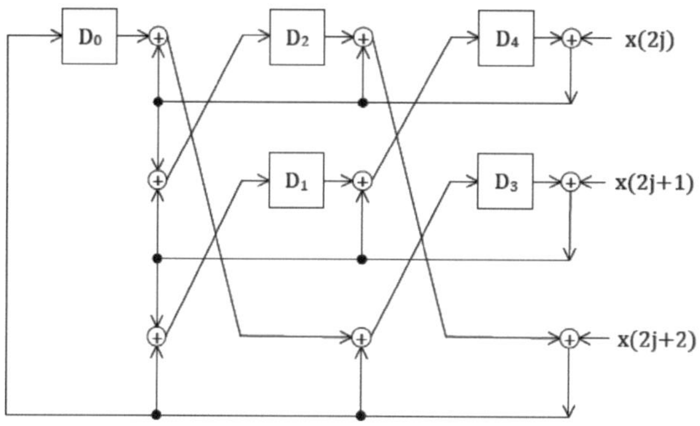

Figure 4.
CRC implementation for $g(x) = x^5 + x^3 + x + 1$.

processed on every clock cycle. Multiple bits can be processed on every clock cycle to speed the CRC calculation. For example, another CRC implementation for $g(x) = x^5 + x^3 + x + 1$ is shown in **Figure 4** [14, 15]. The parallelism of this CRC implementation is 3 and thus three bits are processed on every clock cycle. From **Figures 3** and **4**, it is clear that parallelism comes at the expense of the increased circuit complexity.

4.2 CRC implementation by parallel processing

In this scheme, $\boldsymbol{e}$ is segmented into multiple blocks and the CRC of each block is obtained by parallel processing. $\boldsymbol{e}$ is segmented into multiple blocks [16]:

$$\boldsymbol{e}^0, \boldsymbol{e}^1, \dots, \boldsymbol{e}^{M-1} \tag{11}$$

The size of $\boldsymbol{e}^{M-1}$ is n and the size of $\boldsymbol{e}^j$ is m, where $0 \le j \le M-2$. Note that $L + N$ is equal to $n + m(M-1)$. As a consequence, $\boldsymbol{e}$ can be expressed as

$$\boldsymbol{e} = \boldsymbol{e}^0 x^{(M-2)m+n} + \boldsymbol{e}^1 x^{(M-3)m+n} + \dots + \boldsymbol{e}^{M-3} x^{n+m} + \boldsymbol{e}^{M-2} x^n + \boldsymbol{e}^{M-1}$$

The CRC of $\boldsymbol{e}$ is given by

$$\begin{aligned} \mathrm{CRC}[\boldsymbol{e}] &= \mathrm{CRC}\left[\boldsymbol{e}^0 x^{(M-2)m+n} + \dots + \boldsymbol{e}^{M-2} x^n + \boldsymbol{e}^{M-1}\right] \\ &= \mathrm{CRC}\left[\mathrm{CRC}\left[\boldsymbol{e}^0\right]\mathrm{CRC}\left[x^{(M-2)m+n}\right] + \dots + \mathrm{CRC}\left[\boldsymbol{e}^{M-2}\right]\mathrm{CRC}[x^n]\right] + \mathrm{CRC}\left[\boldsymbol{e}^{M-1}\right] \\ &= \mathrm{CRC}\left[\left[\mathrm{CRC}\left[\boldsymbol{e}^0\right]\mathrm{CRC}\left[x^{(M-2)m}\right] + \dots + \mathrm{CRC}\left[\boldsymbol{e}^{M-2}\right]\right]\mathrm{CRC}[x^n]\right] + \mathrm{CRC}\left[\boldsymbol{e}^{M-1}\right] \end{aligned} \tag{12}$$

The above expression explains how $\mathrm{CRC}[\boldsymbol{e}]$ is obtained. The detail is shown in **Figure 5**. $\mathrm{CRC}[x^m]$, $\mathrm{CRC}[x^{2m}]$, …, $\mathrm{CRC}\left[x^{(M-2)m}\right]$ and $\mathrm{CRC}[x^n]$ do not depend on the transport block size and can be precomputed. Since n is in the range $[0, m-1]$, variables that need to be precomputed include

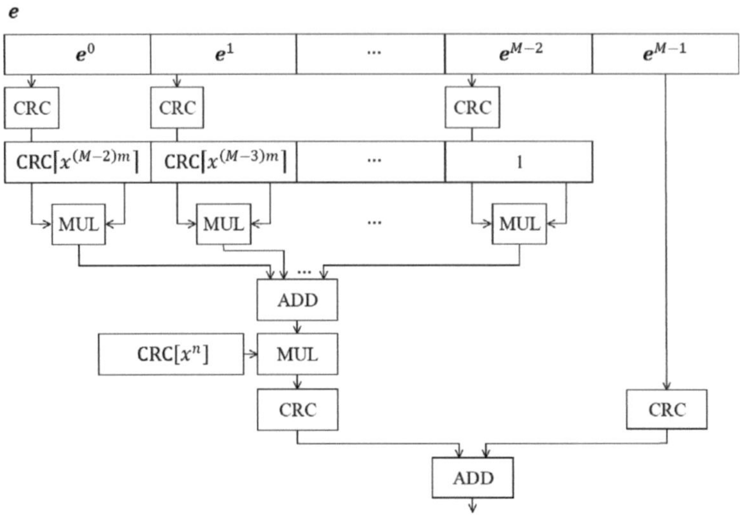

Figure 5.
CRC implementation by parallel processing.

$$\mathrm{CRC}\left[x^1\right], \mathrm{CRC}\left[x^2\right], \dots, \mathrm{CRC}\left[x^{m-1}\right], \mathrm{CRC}[x^m], \mathrm{CRC}\left[x^{2m}\right], \dots, \mathrm{CRC}\left[x^{(M-2)m}\right] \quad (13)$$

As a consequence, the number of variables that needs to be precomputed is $m + M - 3$.

It is clear that the memory that needs to store the variables increases with the transport block size. To reduce the memory, $\mathrm{CRC}[x^{\alpha m}]$ can be recursively calculated by using $\mathrm{CRC}[x^m]$ [17]. That is, $\mathrm{CRC}[x^{\alpha m}]$ is recursively obtained by the following expression

$$\mathrm{CRC}[x^{\alpha m}] = \mathrm{CRC}\left[\mathrm{CRC}\left[x^{(\alpha-1)m}\right]\mathrm{CRC}[x^m]\right] \quad (14)$$

In this way, the variables that need to be precomputed include

$$\mathrm{CRC}\left[x^1\right], \mathrm{CRC}\left[x^2\right], \dots, \mathrm{CRC}\left[x^{m-1}\right], \mathrm{CRC}[x^m] \quad (15)$$

As a consequence, the number of variables that needs to be precomputed is m.

4.3 CRC implementation by serial processing

In this scheme, $\boldsymbol{e}$ is segmented into multiple blocks and the CRC of each block is obtained by serial processing. $\boldsymbol{e}$ is segmented into multiple blocks [18]:

$$\boldsymbol{e}^0, \boldsymbol{e}^1, \dots, \boldsymbol{e}^{M-1} \quad (16)$$

The size of $\boldsymbol{e}^{M-1}$ is n and the size of $\boldsymbol{e}^j$ is m, where $0 \le j \le M - 2$. Note that $L + N$ is equal to $n + m(M - 1)$. $\boldsymbol{e}$ can be expressed as

$$\boldsymbol{e} = \boldsymbol{e}^0 x^{(M-2)m+n} + \boldsymbol{e}^1 x^{(M-3)m+n} + \dots + \boldsymbol{e}^{M-3} x^{n+m} + \boldsymbol{e}^{M-2} x^n + \boldsymbol{e}^{M-1} \quad (17)$$

The CRC of $\boldsymbol{e}$ is given by

$$
\begin{aligned}
T_1 &= \mathrm{CRC}\left[\boldsymbol{e}^0 x^{(p-1)m} + \ldots + \boldsymbol{e}^{1p-1}\right] \\
T_2 &= \mathrm{CRC}\left[\boldsymbol{e}^p x^{(p-1)m} + \ldots + \boldsymbol{e}^{2p-1}\right] + \mathrm{CRC}[T_1 x^{mp}] \\
&\ldots \\
T_e &= \mathrm{CRC}\left[\boldsymbol{e}^{(e-1)p} x^{(p-1)m} + \ldots + \boldsymbol{e}^{ep-1}\right] + \mathrm{CRC}[T_{e-1} x^{mp}] \\
T_{e+1} &= \mathrm{CRC}\left[\boldsymbol{e}^{ep} x^{(M-ep-2)m} + \ldots + \boldsymbol{e}^{M-2}\right] + \mathrm{CRC}\left[T_e x^{(M-eP-1)m}\right] \\
T_{e+2} &= \mathrm{CRC}[T_{e+1} x^n] + \mathrm{CRC}\left[\boldsymbol{e}^{M-1}\right]
\end{aligned}
\tag{18}
$$

where $e = \lfloor (M-1)/P \rfloor$. The above expression explains how $\mathrm{CRC}[\boldsymbol{e}]$ is calculated. The detail is shown in **Figure 6**. $\mathrm{CRC}[x^m]$, $\mathrm{CRC}[x^{2m}]$, ..., $\mathrm{CRC}\left[x^{(p-1)m}\right]$ and $\mathrm{CRC}[x^n]$ do not depend on the transport block size and can be precomputed. Since n is in the range $[0, m-1]$, variables that need to be precomputed include

$$
\mathrm{CRC}\left[x^1\right], \mathrm{CRC}\left[x^2\right], \ldots, \mathrm{CRC}\left[x^{m-1}\right], \mathrm{CRC}[x^m], \mathrm{CRC}\left[x^{2m}\right], \ldots, \mathrm{CRC}\left[x^{(p-1)m}\right] \tag{19}
$$

As a consequence, the number of variables that needs to be precomputed is $m + p - 2$.

It is clear that the memory that needs to store the variables increases with the transport block size. To reduce the memory, $\mathrm{CRC}[x^{am}]$can be recursively calculated by using $\mathrm{CRC}[x^m]$ [17]. That is, $\mathrm{CRC}[x^{am}]$ is recursively obtained by the following expression

$$
\mathrm{CRC}[x^{\alpha m}] = \mathrm{CRC}\left[\mathrm{CRC}\left[x^{(\alpha-1)m}\right]\mathrm{CRC}[x^m]\right] \tag{20}
$$

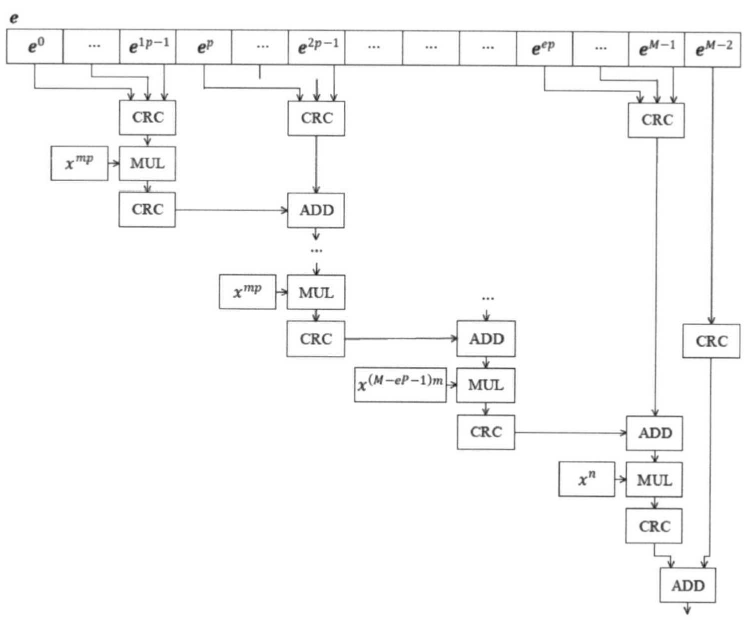

Figure 6.
CRC implementation by serial processing.

In this way, the variables that need to be precomputed include

$$\mathrm{CRC}\left[x^1\right], \mathrm{CRC}\left[x^2\right], \ldots, \mathrm{CRC}\left[x^{m-1}\right], \mathrm{CRC}\left[x^m\right] \tag{21}$$

As a consequence, the number of variables that needs to be precomputed is m.

4.4 The Sarwate algorithm

Sarwate proposes an algorithm based on the lookup table [19]. The detail and the proof of the algorithm can be found in [19]. The Sarwate algorithm is shown in

```
crc = INIT_VALUE;
while (p_buf < p_end)
{
   crc = table[(crc ^ *p_buf++) & 0x000000FF] ^ (crc >> 8);
}
return crc ^ FINAL_VALUE;
```

Figure 7.
The Sarwate algorithm.

```
crc = INIT_VALUE;
while (p_buf < p_end)
{
   crc ^= * (uint32_t *) p_buf;
   term1 = table_56[crc & 0x000000FF] ^
           table_48[(crc >> 8) & 0x000000FF];
   term2 = crc >> 16;
 crc = term 1 ^
         table_40[term2 & 0x000000FF] ^
         table_32[(term2 >> 8) & 0x000000FF];
   p_buf  += 4;
}
return crc ^ FINAL_VALUE;
```

Figure 8.
The slicing-by-4 algorithm.

```
crc = INIT_VALUE;
while (p_buf < p_end)
{
   crc ^= * (uint32_t *) p_buf;
 p_buf += 4;
   term1 = table_88[crc & 0x000000FF] ^
           table_80[(crc >> 8) & 0x000000FF];
   term2 = crc >> 16;
 crc = term 1 ^
         table_72[term2 & 0x000000FF] ^
         table_64[(term2 >> 8) & 0x000000FF];
   term1 = table_56[(*(uint32_t *)p_buf) & 0x000000FF] ^
           table_48[((*(uint32_t *)p_buf) >>8) & 0x000000FF];
   term2 = (*(uint32_t *) p_buf ) >>16;
 crc = crc ^
         term1 ^
         table_40[term2 & 0x000000FF] ^
         table_32[(term2 >> 8) & 0x000000FF];
   p_buf  += 4;
}
return crc ^ FINAL_VALUE;
```

Figure 9.
The slicing-by-8 algorithm.

Figure 7 [20]. The Sarwate algorithm uses a single table of 256 32-bit elements and reads the bits byte by byte. Modern processors usually access 32 bits or 64 bits at a time. As a consequence, the Sarwate algorithm is not efficient. Some schemes have been proposed in the literatures to solve this problem.

4.5 The slicing-by-4 and slicing-by-8 algorithms

Kounavis and Berry propose the slicing-by-4 and slicing-by-8 algorithms based on the lookup table [20]. The detail and the proof of the algorithms can be found in [20]. The slicing-by-4 and slicing-by-8 algorithms are shown in **Figures 8** and **9** respectively [20]. The slicing-by-4 algorithm uses four tables of 256 32-bit elements and reads 32 bits at a time. The slicing-by-8 algorithm uses eight tables of 256 32-bit elements and reads 64 bits at a time. The performance of the slicing-by-4 and slicing-by-8 algorithms is improved compared to the Sarwate algorithm.

5. Conclusion

In 5G NR, the transport block consists of up to million bits and the code block consists of up to 8448 bits. Due to the difference in the size of the transport block and the code block, the scheme of the CRC processing suitable for the transport block and that suitable for the code block are different. This chapter gives an overview of the CRC implementation in 5G NR.

Conflict of interest

The authors declare no conflict of interest.

Author details

Hao Wu[1,2]

1 Department of Wireless Product Research and Design Institute, ZTE Corporation, Shenzhen, China

2 State Key Laboratory of Mobile Network and Mobile Multimedia Technology, ZTE Corporation, Shenzhen, China

*Address all correspondence to: wu.hao19@zte.com.cn

References

[1] Erik D, Stefan P, Johan S. 4G: LTE/LTE-advanced for Mobile Broadband. 2nd ed. Oxford, UK: Elsevier; 2014

[2] Erik D, Stefan P, Johan S. 5G NR: The Next Generation Wireless Access Technology. London, UK: Elsevier; 2018

[3] 3GPP TS 38.211, V15.3.0, NR; Physical channels and modulation (Release 15). 2018-09

[4] Hyoungju J, Sunho P, Jeongho Y, Younsun K, Juho L, Byonghyo S. Ultra-reliable and low-latency communications in 5G downlink: Physical layer aspects. IEEE Wireless Communications. 2018;**25**:124-130. DOI: 10.1109/MWC.2018.1700294

[5] Petar P, Kasper Floe T, Osvaldo S, Giuseppe D. 5G wireless network slicing for eMBB, URLLC, and mMTC: A communication-theoretic view. IEEE Access. 2018;**6**:55765-55779. DOI: 10.1109/ACCESS.2018.2872781

[6] 3GPP TS 38.214, V15.3.0, NR; physical layer procedures for data (Release 15). 2018-09

[7] 3GPP TS 38.212, V15.3.0, NR; multiplexing and channel coding (Release 15). 2018-09

[8] R1-1713458. Qualcomm Incorporated, CRC attachment. 3GPP TSG RAN Meeting #90; August 21–25, 2017; Prague, Czechia

[9] Hao W. Hard decision of the zero a posteriori LLR in 5G NR. Internet Technology Letters. 2020;**3**:e146. DOI: 10.1002/itl2.146

[10] Yan S, Min SK. A table-based algorithm for pipelined CRC calculation. In: Proceedings of the IEEE International Conference on Communications (ICC'10); 23-27 May 2010; Cape Town, South Africa: IEEE; 2010. pp. 1-5

[11] Hao W, Fang W, Yuqing Y. A distributed CRC early termination scheme for high throughput QC-LDPC codes. In: Proceedings of International Conference on Wireless Communications and Signal Processing (WCSP) (WCSP '18); 18-20 October 2018; Hangzhou, China: IEEE; 2018. pp. 1-5

[12] TongBi P, Charles Z. High-speed parallel CRC circuits in VLSI. IEEE Transactions on Communications. 1992; **40**:653-657. DOI: 10.1109/26.141415

[13] Richard EB. Algebraic Codes for Data Transmission. Cambridge, UK: Cambridge University Press; 2003

[14] Chao C, Keshab KP. High-speed parallel CRC implementation based on unfolding, pipelining, and retiming. IEEE Transactions on Circuits and Systems II: Express Briefs. 2006;**53**: 1017-1021. DOI: 10.1109/TCSII.2006.882213

[15] Keshab KP. VLSI Digital Signal Processing Systems: Design and Implementation. New York, USA: John Wiley & Sons; 1999

[16] Ji HM, Killian E. Fast parallel CRC algorithm and implementation on a configurable processor. In: Proceedings of the IEEE International Conference on Communications (ICC'02); 28 April-2 May 2002. New York, USA: IEEE; 2002. pp. 1813-1817

[17] Hyeji K, Injun C, Wooseok B, Jong-yeol L, Ji-Hoon K. Distributed CRC architecture for high-radix parallel turbo decoding in LTE-advanced systems. IEEE Transactions on Circuits and Systems II: Express Briefs. 2015;**62**:

906-910. DOI: 10.1109/TCSII.2015.
2435131

[18] Hao W, Tao L, Jin X, Fang W. Parallel CRC architecture for broadband communication systems. Electronics Letters. 2017;**53**:1439-1441. DOI: 10.1049/el.2017.1029

[19] Sarwate DV. Computation of cyclic redundancy checks via table look-up. Communications of the ACM. 1988;**31**: 1008-1014. DOI: 10.1145/63030.63037

[20] Michael EK, Frank LB. Novel table lookup-based algorithms for high-performance CRC generation. IEEE Transactions on Computers. 2008;**57**: 1550-1560. DOI: 10.1109/TC.2008.85

Section 2

Practical Aspects from Next Generation Mobile Landscape

Chapter 5

Prospects of 5G Satellite Networks Development

Valery Tikhvinskiy and Victor Koval

Abstract

In the future, 5G networks will represent the global telecommunication infrastructure of the digital economy, which should cover the whole world including inaccessible areas not covered by 5G terrestrial networks. Given this, the satellite segment of 5G networks becomes one of the pressing issues of development and standardization at the second stage of 5G networks development in the period 2020–2025. The requirements for 5G satellite network will be determined primarily by combination of key services supported by 5G networks, which are combined by three basic business models of 5G terrestrial networks: enhanced Mobile Broadband Access (eMBB), Massive Internet of Things connections (mIoT), and Ultra-reliable low-latency communication (uRLLC). 3GPP as leading international standards body has identified several use cases and scenarios of 5G satellite networks development. 5G satellite networks are understood to mean networks in which the NG-RAN radio access network is constructed using a satellite network technology. The chapter has discussed the spectral and technological aspects of 5G satellite network developments, issues of architecture and role of delays on quality of services of 5G satellite segment, and possibility of constructing a 5G satellite segment based on distributed and centralized gNB base stations. The issues of satellite payload utilization have considered for bent-pipe and on-board processing technologies in 5G satellite segment.

Keywords: 5G satellite segment, WRC-19, 3GPP, 5G core, gNB, SRI, ISL

1. Introduction

Rapid development and unification of the terrestrial part pertained to IMT-2020 (5G) as well as the limitations claimed for global coverage by terrestrial 5G networks, when using millimeter-wave band (MMWB), calls design engineers for special attention to this potential market segment of mobile satellite telecommunications.

In Summer 2017, within the framework of Paris air show in Le Bourget, European Space Agency (ESA) launched its new project "Satellite for 5G," compiling 16 satellite businesses into consortium aimed at study to introduce 5G satellite-based access components [1].

The consortium is made up of such organizations and institutions as: EURESCOM, Fraunhofer Fokus, Fraunhofer IIS, NewTEC, SES, TU Berlin, and Universität der Bundeswehr. They all have conducted the work concerning the design of SATis5 intended to facilitate implementation, deployment, and evaluation of the integrated 5G satellite network, unveiling the advantages of satellite and terrestrial framework integration for advancing new technologies.

IntechOpen

Additionally, the Working Group FM44 of ECC CEPT completed the preparation of ECC Report "Satellite Solutions for 5G" [2] that will determine the role of satellite component within 5G conception of in relation to the Regions, where services cannot be carried out in circumvention of satellites. In its turn, CEPT came forward with initiative to estimate the pros of satellites for 5G in terms of efficiency, capacity, and stability. Since CEPT Administrations are considering the issues related to 5G implementation in the nearest future, so the studies of satellite access in 5G are expected to facilitate the process of decision-making regarding the potential role of satellite subscriber's links in the 5G ecosystem.

2. Concept of 5G satellite implementation

The implementation of 5G satellite component for 5G service access on principles "at any time with any user in any place" helps to meet many challenges. However, there are numerous other hindrances requiring comprehensive and global studies.

Generally, the regions that are subject to coverage by terrestrial mobile networks of radio access are of fragmentary nature and correspond with the places of population concentration, regarding the economic expediency of base stations building. In some cases, the sparsely populated territories not covered by modern telecommunications. Thus, at the outset of 3G (IMT-2000) development, the universal coverage by mobile services was the key prerequisite for network construction, contributing to the formation of global 3G segment. However, in the course of 4G network evolution, the idea of global coverage by these networks was not even contemplated, in the hope of finding the convergent solutions in the field of satellite and terrestrial mobile telecommunications.

The concept of 5G satellite component considered nowadays rests upon the following preconditions [3]:

- 5G satellite component is to be integrated into other mobile and fixed networks, but not as autonomy one for the provision of 5G services. The integration of satellite and terrestrial 5G segments forms the key aspect of this vision;

- Satellite communication systems are fundamental components for reliable delivering of mobile services, not only in Europe, but also in other regions of the world as by continuum over time and at a reasonable price;

- 5G satellite component will facilitate universality of 5G networks as well as the solution for various issues dealt with maintenance of multimedia traffic growth, global coverage, M2M, and critical telecommunications (emergency and natural disasters) in optimizing costs for end-users;

- Satellite component may become a part of configuration for 5G hybrid network, consisted of combination of broadcasting and broadband infrastructures, run in a manner to ensure uninterrupted and online convergence of 5G services for all end-users.

The requirements for 5G satellite component will be defined by, first of all, the aggregate of services carried out by 5G, consolidated in the families of usage scenarios of 5G terrestrial segment [4, 5]: enhanced Mobile Broadband (eMBB), Massive Machine-Type Communications (mMTC or mIoT), and Ultra-Reliable Low Latency Communication (URLLC).

The potential of satellite networks to uphold the key scenarios for 5G applications is specified by already existing characteristics applicable to modern satellite networks as well as tendencies in satellite technology development in future:

- *eMBB scenario*. According to this scenario, satellite networks are capable of maintaining data transfer at speed up to several gigabits per second, meeting the requirements for extended services of mobile broadband eMBB. Nowadays, satellite technologies can broadcast thousands of channels with the content of high bandwidth (HD and UHD). In its turn, this potential can be used to support the mobile network services of future generation. At present, satellites are being used as transport networks within 2G/3G in many regions of the world, whereas high-throughput satellites (HTS) of modern and future generations on geostationary and non-geostationary orbits can maintain transport infrastructure of mobile networks 4G/LTE and 5G in future.

- *mMTC scenario*. Satellite communication systems are already keeping up the technology of SCADA and other global applications for cargo and object tracking in the context of IoT devices mass use. Their capabilities can be scaled up to support devices and services of IoT within the direct control channel or as a feedback line with IoT and M2M devices from remote locations, ships, and other carrying vessels.

- *uRLLC scenario*. Satellite communication systems gained notoriety owing to its and their satellite communication systems gained notoriety by owing to its and their ability to meet the case concerning the requirements for network signal delays, aiming at procuring critical and highly reliable communications. The principal users of these networks are international broadcasters, mobile network operators, governmental bodies, and commercial users. The applications that turn out to be more sensitive to signal delays can be bolstered via new medium and low earth orbit satellite networks, which will to be deployed.

5G satellite networks are such networks, where radio access network NG-RAN is designed by means of satellite network utilization. Technical specifications of 3GPP [3] identified several cases for 5G satellite network use, presented below.

Case 1. Roaming between terrestrial and satellite networks. In this case, 5G satellite network operator provides data services delivery on globally coverage basis. An operator of terrestrial 5G network, in its turn, concludes roaming agreement with the operator of 5G satellite network operator as well as the other terrestrial network operators. User terminal exploits 5G satellite network only in the absence of radio coverage by terrestrial 5G networks.

Case 2. Broadcast and multicast with a satellite overlay. In this case, the operator of 5G satellite network provides video broadcasting or any other delivery of services within the global territory. The existing terrestrial mobile networks, supplying broadcasting services, can rely on 5G satellite network aiming at meeting its primary objectives related to the expansion of radio resource, broadcasting content, and ensuring global access to content.

Case 3. Internet of Things with a satellite network. In this case, 5G satellite network operator provides the delivery of IoT-services globally. Space segment of 5G satellite network uses low-orbiting satellites so as to ensure radio connections for IoT devices with low power consumption.

Case 4. Temporary use of a satellite component. In this case, a number of 5G network operators with access to the satellite component grant access to their network with a minimum set of service (such as voice, messaging, and mail) so as to provide to each user devices under the satellite coverage a guaranteed access.

Case 5. Optimal routing or steering over a satellite. The 5G networks will combine available terrestrial and satellite network components to optimize the connectivity of user devices in accordance with the requested QoS. Depending on the quality requirements to QoS-parameter 5QI as well as bandwidth, the optimal traffic routing is secured within the territories of joint radio coverage (of satellite and terrestrial networks). In a 5G network with satellite access, user devices with terrestrial access and supporting satellite networks access will be capable of dual connectivity with a satellite access network and a terrestrial access network. A 5G network with satellite access will be capable of establishing independently uplink and downlink connectivity through the 5G satellite and 5G terrestrial access networks.

Case 6. Satellite transboundary service continuity. This case provides for 5G global satellite network within the territory of a few countries. According to the prerequisites established by legislation of the relevant states, subscribers' traffic is to be terminated in user location, within the licensed network. Consequently, in compliance with this statement, 5G satellite network is being designed as access network to respective terrestrial networks, covering the territories of various states. Therefore, it can also be used as autonomous 5G network on neutral territories.

Case 7. Global satellite overlay. In this case, global low-orbiting satellite network will be utilized as the overlaying network of terrestrial data network. The topology of communication links will be defined on basis of minimizing delivery time of protocol data unit. Thus, the main idea considers that delay of signal propagation equals the speed of light (299,792,458 m/s) in airspace, whereas in optical fiber, this parameter achieves up to 2/3 of speed of light. Based on the above, time duration equals 1 ms correlates with propagation distance of 300 km in airspace and 200 km in optical fiber (excluding curvature of circuit). With more large distance between the source and recipient of a message (reaching several thousand km), the difference in time delivery may be significant and actually for a series of applications in banking, burs exchange, and industry fields.

Case 8. Indirect connection through a 5G satellite access network. This case will be assumed that mass user devices will be deprived access to satellite interface. Interaction of these 5G user devices with satellite networks is carried out through relay user units (Relay UE), supporting satellite interface. This relay UEs can function separately or will be set into rescue vessels, air planes, and railway carriages. While implementing these indirect connections of 5G user devices through satellite access networks, it is vital to solve the issues dealing with security, tariffing, etc.

Case 9. 5G fixed Backhaul between NR and the 5G core. This option considers the use of satellite network by organizations of transport channels Midhaul, Backhaul between stationary base stations gNB and 5G core network. The interfaces between the 5G core and NR are transported directly over the satellite link.

Case 10. 5G Moving platform Backhaul. This case considers the utilization of satellite network for transport link organizations in 5G network (Moving Platform) such as Midhaul, Backhaul between moving gNBs and 5G core network. Moving 5G base stations can be placed on river and maritime vessels, trains, etc.

Case 11. 5G to premises. This case implies that 5G satellite network interoperates with non-3GPP technologies (for instance, IEEE 802.11, IEEE 802.16). It is using a home/office gateway unit to combine the available signals from 5G satellite network and to present modern Wi-Fi coverage within the premises.

Case 12. Satellite connection of remote service center to off-shore wind farm. In this use, case 5G satellite network based on Low Earth Orbit (LEO) satellite used for set up satellite link connection with local control center in the wind power

Types of satellite orbits	Height, km	Number of satellites
Low earth orbit (LEO)	800	≈ 80
	1400	≈ 50
Medium earth orbit (MEO)	8000	≈ 10
Geostationary earth orbit (GEO)	35,786	≈ 3

Table 1.
Minimum satellites needed to maintain global radio coverage.

Types of satellite orbits	Delays in link "User terminal-satellite", ms		Maximum one-way delay, ms
	Minimum	Maximum	
LEO	3	15	30
MEO	27	43	90
GEO	120	140	280

Table 2.
UE to satellite propagation delay.

plant communication network includes a 5G satellite user device. It will be provided low satellite communication latency and high uplink/down data transmission volume.

However, these cases do not finish and limit possibility of 5G satellite segment applications and will be proceeded in 3GPP study in Release 17 on 5G evolution.

The main flaw of satellite segment consists in increased delay of information transfer owning to distance between user units and gNB base station. The requirements submitted to the quality of service for data transfer within 5G satellite segment also depend on the relevant number of satellites in operation. The minimum quantity of satellites in operation needed to maintain radio coverage for orbits of different heights [6] is shown in **Table 1**.

The signal delays forming for different satellite orbits and satellite limits on satellite segment delays are presented in **Table 2**. Additionally, the indicated delays are summarized with 5 ms delay, added by satellite. Therefore, maximum delay limits reach 30, 90, and 280 ms.

Other QoS-requirements (Default Priority Level, Packet Delay Budget, Packet Error Rate) for 5G satellite segment have set in 3GPP technical specifications.

3. Spectrum aspects of 5G satellite segment use

On the one hand, spectrum and wide bandwidth for 5G terrestrial networks will require utilization of millimeter-wave (mm-wave) bands to provide data transfer speed reaching up to 20 Gigabits per second in 5G radio interface connect with the process of delivery of the extended broadband mobile access (eMBB) service. On other hand such requirements to use frequency channels with bandwidth from 50 up to 400 MHz for eMBB-services can provide only in mm-wave bands which already utilized within satellite networks. That is why mm-wave bands in nearest future will turn out to be the most requested in 5G and satellite communications.

World Radiocommunication Conference 2019 (WRC-19) allocated of additional mm-wave frequency bands 24.25–27.5 GHz, 37–43.5 GHz, and 66–71 GHz for 5G terrestrial networks on a global basis. In a series of countries and regions, frequency bands of 45.5–47 GHz and 47.2-48.2 GHz received complimented allocation to terrestrial segment of IMT. This decision WRC-19 will be allowed to use some part of mm-wave bands on spectrum sharing basis for 5G satellite and 5G terrestrial network segments.

Table 3 shown the basic frequency bands allocated to fixed and mobile satellite services, sited within the band from 10.7 to 275 GHz, designed for satellite networks and satisfied the needs for 5G channel bandwidths [7].

The analysis of spectrum bands within 12.5–86 GHz has revealed the availability of frequency bands with total bandwidth equals 17.75 GHz in up-link (UL) bands and within 10.7–76 GHz – the availability of frequency bands with total bandwidth equals 20 GHz in down-link (DL) bands for satellite networks.

In order to ensure the provision of services in the field of mass deployment of IoT devices in 5G satellite segment, it was suggested that part of S-band should utilize as a potential option with 30 MHz bandwidth [8]:

- uplink (IoT device–satellite) in band: 1980–2010 MHz;
- downlink (satellite–IoT device) in band: 2170–2200 MHz.

The connection between satellite 5G base station gNB and feeder link of satellite network can be performed in one of the fixed satellite service bands.

Furthermore, the study of most popular frequency bands, namely Ka-band (28 GHz) and Q/V-bands (37–53 GHz), has exposed the following features which are to be considered while elaborating the solutions for 5G.

While considering the use of Ka-band for 5G satellite segment, one should bear in mind that:

- Ka-band is a traditional satellite band, enhancing access for satellite networks;
- a part of this band has allocated for 5G terrestrial networks on a global basis by WRC-19;
- a few national administrations are reviewing this band in terms of 5G terrestrial networks use.

While considering the use of Q/V-bands (37-53 GHz) for 5G satellite network, one should bear in mind that:

- V-band has not been used yet for satellite applications, in particular, for feeder lines of satellite network;
- a part of V-band has been added into bands which has allocated for 5G terrestrial networks on a global basis by WRC-19;
- 3GPP accelerates common efforts on joint researches as well as study of requirements attached to satellite as well as terrestrial segment of 5G in V-band in Release 17.

Thus, 5G satellite segment can be constructed as the multiband one, as well as 5G terrestrial segment, which was divided into frequency bands lower 6 GHz (FR1) and higher 6 GHz (FR2) also.

Up-link		Down-link		Intersatellite link	
Frequency range (GHz)	**Bandwidth (GHz)**	**Frequency range (GHz)**	**Bandwidth (GHz)**	**Frequency range (GHz)**	**Bandwidth (GHz)**
12.5–13.25	0.75	10.7–11.7	1.0	22.55-23.55	1.0
13.75–14.8	1.0	17.7–21.2	3.5	25.25-27.5	2.25
27.5–31.0	3.5	37.0–42.5	5.5	59.0-66.0	7.0
42.5–47.0	4.5	66.0-76.0	10.0	66.0-71.0	5.0
48.2–50.2	2.0	123.0-130.0	7.0	116.0-123.0	7.0
50.4–51.4	1.0	158.5-164.0	5.5	130.0-134.0	4.0
81.0–86.0	5.0	167.0-174.5	7.5	174.5-182.0	7.5
209.0–226.0	17.0	191.8-200.0	8.2	185.0-190.0	5.0
252.0–275.0	23.0	232.0-240.0	8.0		
Total of bandwidth	**57.75**	Total of bandwidth	**56.2**	Total of bandwidth	**38.75**

Table 3.
Frequency bands allocated to fixed and mobile satellite services.

4. Satellite segment architecture for 5G networks

The main standardization body – 3GPP responsible for technical specifications on 5G equipment and 5G infrastructure conducted first studies regarding 5G satellite segment use, while elaborating Release 14 within Technical report 3GPP TR 38913 [4].

5G satellite options, presented by 3GPP related to the deployment of 5G satellite segment, are designed for 5G services delivery in areas, where their provision by 5G terrestrial segment is impeded as well as for the services supported by satellite systems.

According to Report [4], 5G satellite segment is to complement 5G services, which delivering especially on road, rail and waterways as well as in rural regions, where access to 5G terrestrial segment is unavailable. 5G services supported via 5G satellite segment go beyond data and voice communications, providing connection with IoT devices and M2M, access to broadcasting services and a number of other services, that is tolerant of signal delays.

Partnership project 3GPP has come up with three options in respect of deployment, shown in **Table 4** [4].

The satellite orbits, shown in **Table 4** and in **Figure 1** enable using:

- Geostationary satellites (GEO), located at an altitude of 35,786 km, providing full coverage of the Earth by a constellation ranging from one up to three satellites between 70°N and 70°S;

- Medium Earth orbit (MEO), located at an altitude of 8000–20,000 km over the surface of the Earth, providing full coverage of the Earth by satellites ranging from 10 up to 12 satellites.

- Low Earth Orbits (LEO) at an altitude of 500–2000 km above the Earth secures the continuity of coverage by satellite network with satellites ranging from 50 up to 100 satellites.

Technical parameters	Option 1	Option 2	Option 3
Carrier frequency	Around 1.5 or 2 GHz for both DL and UL	Around 20 GHz for DL Around 30 GHz for UL	Around 40 or 50 GHz
Duplexing	FDD	FDD	FDD
Satellite architecture	Bent-pipe	Bent-pipe, on-board processing	Bent-pipe, on-board processing
Typical satellite system positioning in the 5G architecture	Access network	Backhaul network	Backhaul network
System bandwidth (DL + UL)	Up to 2 × 10 MHz	Up to 2 × 250 MHz	Up to 2 × 1000 MHz
Satellite orbit	GEO, LEO	LEO, MEO, GEO	LEO, MEO, GEO
UE distribution	100% out-of-doors	100% out-of-doors	100% out-of-doors
UE mobility	Fixed, portable, mobile	Fixed, portable, mobile	Fixed, portable, mobile

Table 4.
Satellites and frequency band options for 5G deployment.

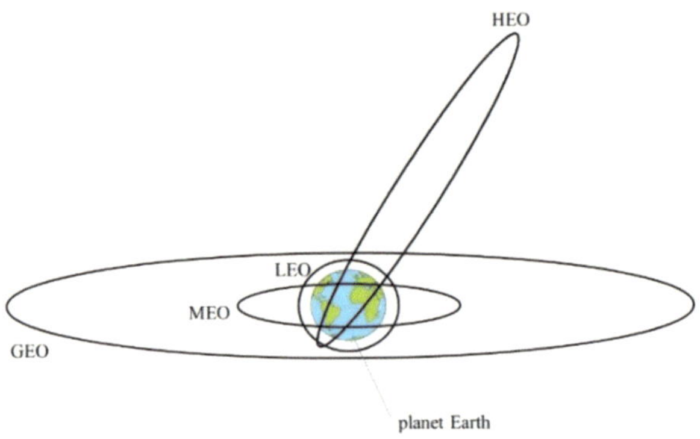

Figure 1.
Typical earth orbit of communication satellite.

The frequency bands, specified in **Table 4**, are applicable solely to a part of satellite bands (**Figure 1**), whereas modern satellite networks are deployed in broad spectrum of frequency bands, including L-band (1–2 GHz), S-band (2–4 GHz), C-band (3.4–6.725 GHz), Ku-band (10.7–14.8 GHz), Ka-band (17.3–21.2 GHz, 27.0–31.0 GHz), Q/V-bands (37.5–43.5 GHz, 47.2–50.2 GHz and 50.4–51.4 GHz), and higher.

The system architecture of 5G satellite segment is being constructed based on the use cases, mentioned in Section 1 of this chapter and two satellite technologies:

1. The architecture based on the technology of bent-pipe (with invisible satellite transponders without On-Board Processing) – this option envisages signal reception from user devices, its amplification, its transfer on other frequency and relaying in the direction of satellite gateway.

2. The architecture based on the technology of On-Board Processing (with satellite transponders, complimented with data processing on board) – this

option implies signal reception from user devices, its regenerations, including modulation and demodulation, encryption and decryption of these signals. The architecture on-board processing also provides for the partial allocation of base station equipment on the board of a satellite.

In December 2017, 3GPP in scope of work on Release 16 was published Report on using satellite access in 5G [3]. The Report submitted new business cases of 5G satellite segment utilization, including Internet of Things alongside with the requirements for performance of cross-border connections and the key characteristics for satellite segment of 5G: types of orbits, coverage area, and signal delays during propagation, network architecture for 5G satellite segment.

In accordance of proposed solutions, 5G satellite segment is inculcated into the integrated radio access network (5G RAN), which will be used satellite infrastructure and 5G core network (5G Core). 5G core can be linked up with the other generation RANs, in particular, 4G RAN, apart from satellite segment for 5G.

System architecture of 5G satellite segment, which is to be set up in accordance with the technology of bent-pipe (with transparent satellite transponders) when signal use solely to amplification and signal conditioning on retention of a modulation type has shown in **Figures 2** and **3**.

As one can see in **Figures 2** and **3**, bent-pipe architecture refers to the architecture where the satellite transponders are transparent: only amplify and change frequency but preserve 5G waveform.

One of the important features of 5G radio access network design is that gNB base stations have a distributed architecture (**Figure 4**) and consist of a central module gNB-CU and one or more distributed modules gNB-DU(s) [9].

The gNB-CU and gNB-DU modules are connected by a logical interface F1. The distributed module gNB-DU supports one or more cells and can only be attached to one central module gNB-CU. This architecture of the gNB base station allows to implement the concept of building an integrated 5G radio access network by placing the gNB-CU and gNB-DU modules at earth stations and realization of F1-interface as a space link based on bent-pipe technology.

System architecture of 5G satellite segment when gNB-CU and gNB-DU modules connected each other through F1-interface by satellite links for on bent-pipe technology has shown in **Figure 5**.

Next options of bent-pipe architecture of 5G satellite segment has used for retranslation NG1 and NG2 interfaces, which connecting 5G base stations gNBs to 5G core. This architecture of 5G satellite segment is shown in **Figure 6**.

In case where 5G user device (UE) has opportunity to use satellite modem with non-3GPP radio interface for bent-pipe architecture of 5G satellite segment, the architecture option of such segment could design as shown in **Figure 7**. 5G satellite segment architecture shall support different configurations where the radio access network is either a satellite NG-RAN or a non-3GPP satellite access network, or both.

Figure 8 shows the 5G satellite segment system architecture implemented on the basis of on-board signal processing technology (with partial deployment of base

Figure 2.
Signals relay architecture for 5G NR radio interface.

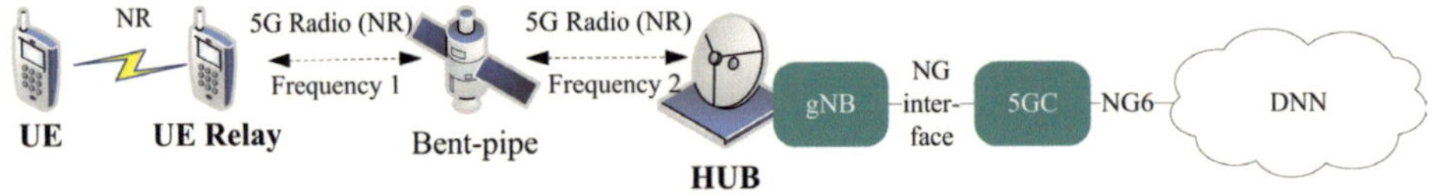

Figure 3.
Relaying architecture based on 5G user device with UE relay.

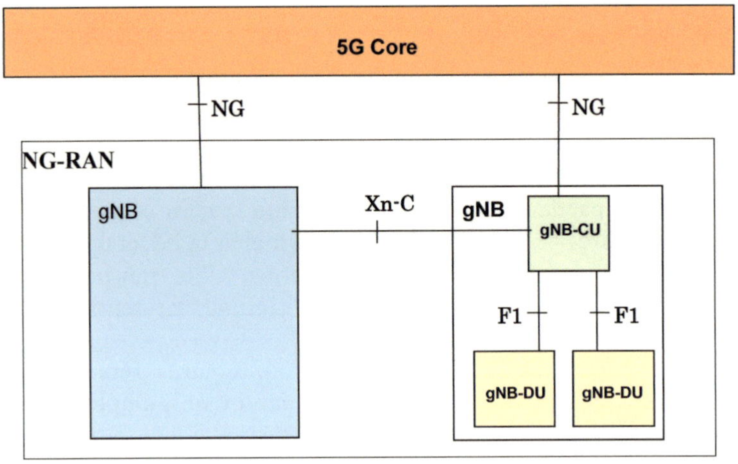

Figure 4.
Architecture 5G base station gNB.

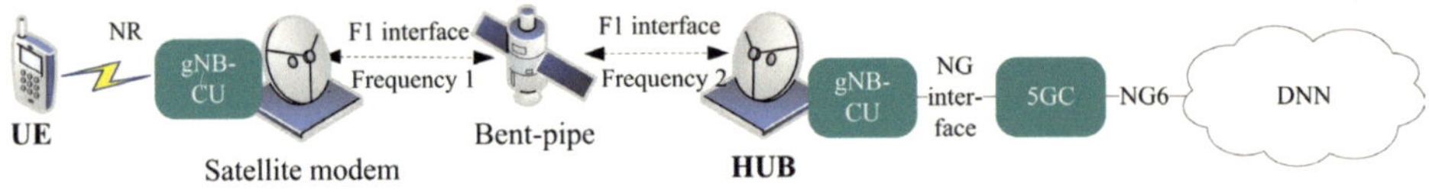

Figure 5.
Architecture 5G base station gNB with F1-satellite interface.

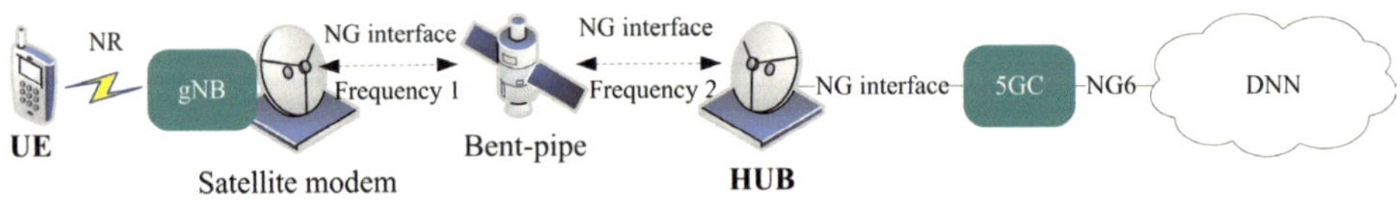

Figure 6.
Signals relay architecture for NG1 and NG2 interfaces.

station processing equipment in satellite). As on-board signal processing payload uses distributed module gNB-DU of 5G base station and as satellite link utilizes 5G NR radio interface.

In accordance design principle of base stations gNBs, some distributed modules gNB-DUs can connect to only one central module gNB-CU. That makes easier 5G coverage of big areas. The solution for 5G satellite segment architecture on regenerative payload enabled NR-RAN with intersatellite links (ISL) for regional or global coverage shown in **Figure 9** [10]. Intersatellite links provide logical F1-interface between distributed modules gNB-DUs, which use Satellite Radio Interface (SRI) over F1 as a transport link between remote radio unit with gNB-CU and satellites.

Second solution for 5G satellite segment architecture (**Figure 10**) has used 5G base station gNB on satellite (as regenerative payload) enabled NR-RAN with ISLs that provide SRI application over Xn-C and Xn-U interfaces. In this case between remote radio units and satellite gNBs will be used, and 5G standard NG-interfaces connect these gNBs with 5G core network.

Mobile devices of 5G satellite segment architecture (**Figures 2–10**) will be presented on the market by user terminals as well as the other wearable devices, installed in cars, ships, planes, etc. Nowadays the potential of wearable satellite user

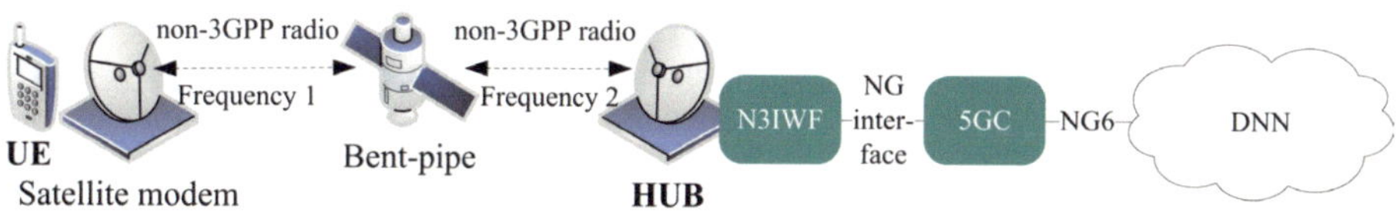

Figure 7.
Signals relay architecture for non-3GPP interface.

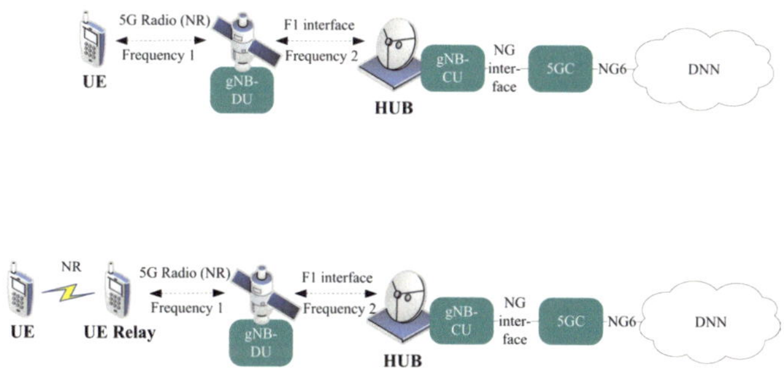

Figure 8.
5G satellite segment architecture based on the on-board processing technology [4].

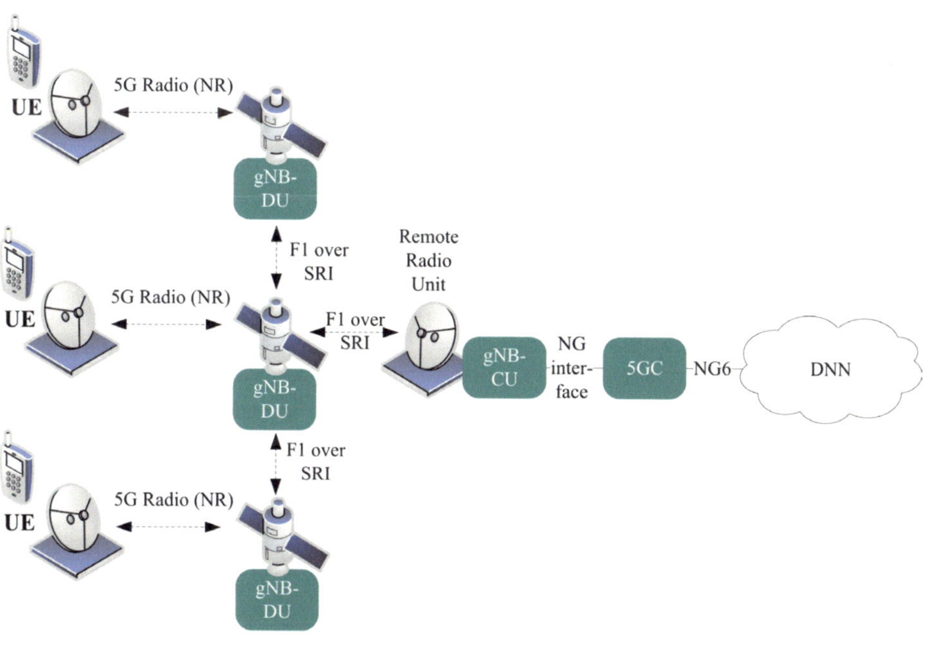

Figure 9.
5G satellite segment architecture on regenerative satellite payloads enabled NR-RAN, with ISL for regional or global coverage.

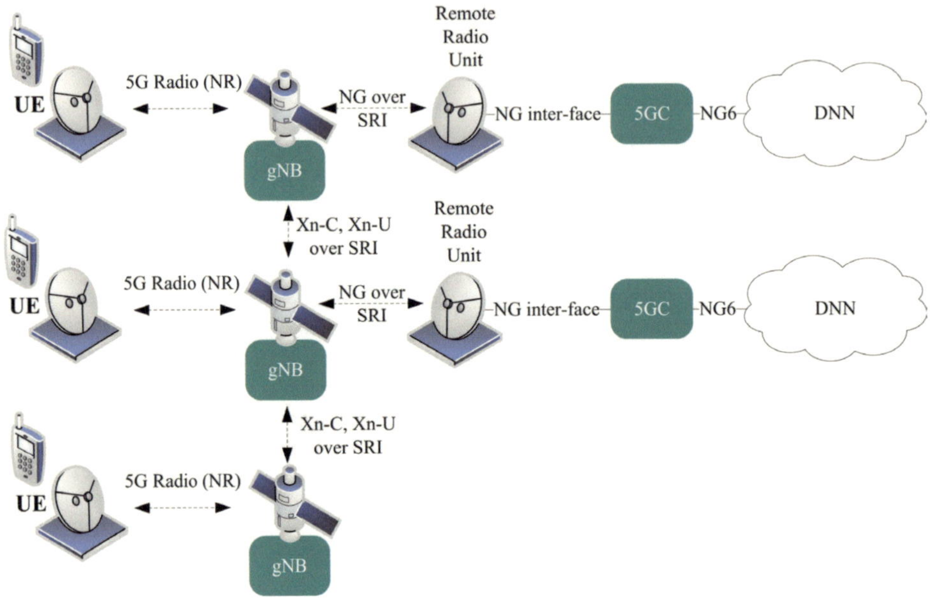

Figure 10.
5GS with regenerative satellite enabled NR-RAN, with ISL and multiple 5G Core connectivity.

terminals is limited to L- and S-frequency bands. However, the studies regarding the potential functioning of 5G satellite user terminals within Ku and mm-wave bands are still ongoing.

5. Projects of the leading manufacturers and researchers

Analysis of proposals and technological projects launched by leading manufacturers and related to usage of satellite networks for expanding the capabilities of 5G networks shows that Boeing [11] and Samsung [7] companies have already tried to make presentations of their projects applicable for 5G satellite segment deployment.

The Boeing company requested the US Federal Communications Commission for permission to launch and operate fixed satellite service (FSS) network on non-geostationary orbit (NGSO). The network would operate in a low-Earth orbit (LEO) in the frequency band 37.5–42.5 GHz (space-Earth) and in the frequency bands 47.2–50.2 and 50.4–52.4 GHz of V-band (Earth-space); it would be used as a NGSO system providing solution of 5G satellite segment operation issues.

The Boeing proposed NGSO system as depicted in **Figure 11** and considered as a 5G satellite segment that is designed to provide a wide range of modern telecommunication services alongside with 5G internet services for a broad types of V-band earth stations and user terminals. V-band user terminals use modern antenna arrays for transmitting and receiving broadband signals in channels of different pass bands. It is to note that a high throughput is supported by multichannel and multiple polarization terminals.

The Boeing presented NGSO system would consist of 2956 LEO satellites for the fixed satellite service network providing high throughput low latency access for user terminals connected through gateway ("hubs") access to 5G network and to a terrestrial optic-fiber network as backhaul connecting to 5G.

The system gateways are expected to be located outside the densely populated areas in the regions with relatively low consumer demand for 5G services. Each

NGSO satellite would form beams, corresponding to cell diameter from 8 up to 11 km on the Earth surface within the overall satellite coverage area.

The NGSO system gateways would operate in the same V-band as user terminals. These gateways would support both frequency and polarization selection of signals with two types of antennas polarization LHCP (Left Hand Circular Polarized) and RHCP (Right Hand Circular Polarized). In addition, the access gateways may contain more than one antenna thereby providing simultaneous access to multiple NGSO satellites visible from a relevant access gateway.

At the first stage of deployment, the Boeing NGSO system would comprise a constellation of 1396 LEO satellites in an altitude of 1200 km. The initial satellite constellation would consist of 35 circular orbital planes with an inclination of 45° and additional 6 circular planes inclined at 55°.

The NGSO system payload (**Figure 12**) would use the improved space-time processing in the course of antenna beam-forming as well as on board digital

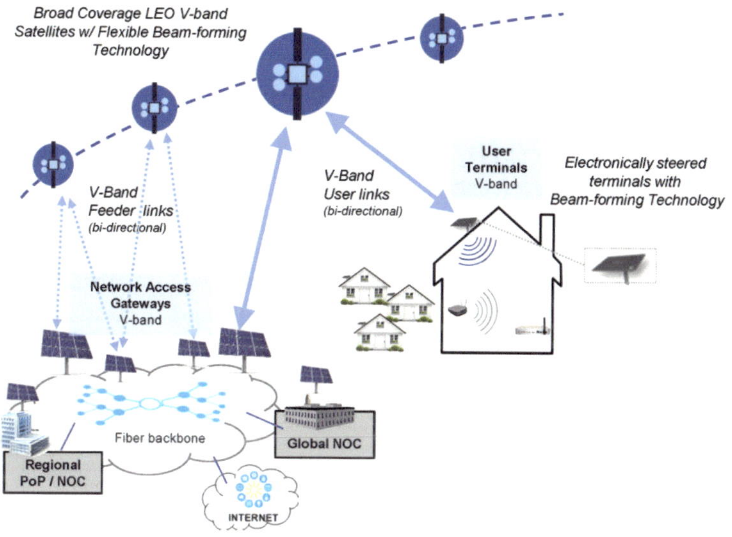

Figure 11.
Satellite solution of the Boeing company.

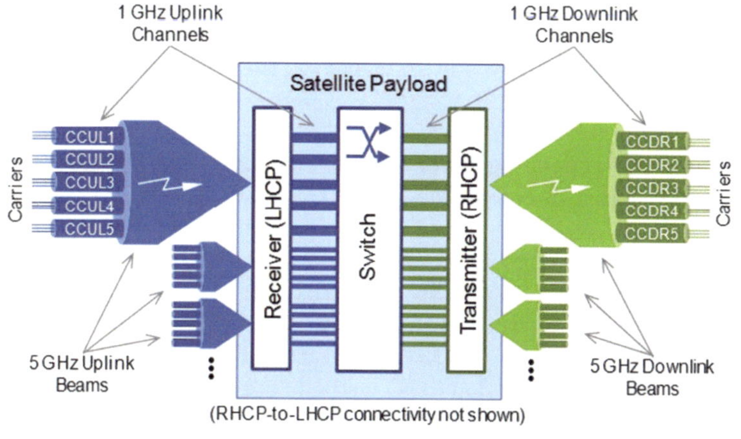

Figure 12.
Scheme of on-board processing payload.

processing so as to generate thousands of narrow-band beams to provide 5G network services through satellite segment on the Earth surface.

Each satellite up-link or down-link may consist of up to five channels of 1 GHz pass band resulting in a total pass band of 5 GHz depending on instant capacity required for a cell supported by a relevant satellite antenna beam. Any satellite UL-channel may be connected to any satellite DL-channel in compliance with used connection algorithm.

Boeing company estimation results show that usage of a satellite network for fixed satellite channels and its spectrum sharing with a 5G terrestrial network in the frequency band 37.5–40.0 GHz would be feasible under the following conditions:

- the frequency band 37.5–40.0 GHz is used only for signal reception in FSS network downlink;

- spectrum sharing between 5G satellite segment and 5G terrestrial segment is feasible due to high satellite elevation angles;

- applying of space-time selection beam-forming methods for terminal antennas of satellite networks and 5G equipment in the aim to achieve higher data rate.

The power flux density (PFD) limits approved by ITU [11, 12] would provide protection for 5G network terrestrial segment from interference caused by FSS satellite network downlinks subject to meeting the requirement of minimal reducing of 5G terrestrial network signal level to 0.2–0.6 dBW.

Boeing simulation results also show that in the assumed spectrum sharing scenario the increasing of 5G base station power would result in enlarging a number of satellite receivers affected by interference from 5G users. Hence, it is required to adopt a (>50 dBW) level of mitigating the interference from 5G networks between FSS earth station receivers and transmitting mobile and base stations of 5G terrestrial segment.

The results of Boeing statistical simulation and quantitative estimation of interference levels show that:

- satellite earth stations may be in a higher degree directly affected by 5G base stations interferences;

- EIRP values for 5G terrestrial segment should be limited to 62–65 dBW, so as to facilitate interference-free shared operation of 5G satellite and terrestrial segments of FSS system to provide for achieving the required data transfer rates in 5G networks.

Therefore, the joint deployment of satellite and terrestrial segments of 5G network is subject to particular conditions related to joint use of spectrum in V-band.

As confirmation, possibility of successfully utilization integrated satellite segment into 3GPP 5G testbed networks was the last demonstration of Surrey University achievements in 5G satellite network development [13].

Three use cases were demonstrated over a live satellite network via Avanti's GEO HYLAS 4 satellite and using iDirect's 5G-enabled Intelligent Gateway (IGW) satellite ground infrastructure that to 5G testbed core network of the University of Surrey to 5G UE terminals. All the 5G testbed use cases used this integrated 5G satellite system for the live satellite connectivity.

The use-case for 5G moving platform was demonstrated over SES's O3b MEO satellite system, using real terminals and 5G core network.

DOI: http://dx.doi.org/10.5772/intechopen.90943

6. Conclusion

The need to provide the coverage of large areas of developed countries with 5G networks and the creation of 5G satellite segment of integrate 5G system become relevant issues of development and standardization of 5G networks at the second stage of building these networks in the period 2020–2025, playing the pivotal role in forging Digital economy.

3GPP efforts allowed to obtain many different use cases of 5G satellite segment applications, architecture solutions on bent-pipe, and on-board processing technologies, which would implement in development of future satellite systems.

The leading international organizations in the field of telecommunications as ITU, 3GPP, 5G PPP joined their efforts with consortiums and satellite manufacturers in conducting the researches related to the elaboration of 5G within the radio frequency ranges that have been allotted to satellite radio service to 5G on WRC-19, especially in S-, Ka- and V-bands.

One of the most important issues of 5G satellite segment future development may refer to shared spectrum usage in the frequency bands allocated to 5G satellite and terrestrial segments on the primary basis. Also urgent is the issue of intersystem electromagnetic compatibility of aboard equipment and earth stations with base stations and user devices of 5G terrestrial segment.

Author details

Valery Tikhvinskiy[1,2]* and Victor Koval[3]

1 Radio Research and Development Institute (NIIR), Moscow, Russian Federation

2 Bauman Moscow State Technical University, Moscow, Russian Federation

3 Geyser-Telecom Ltd., Moscow, Russian Federation

*Address all correspondence to: vtniir@mail.ru

References

[1] The European Space Agency Will Promote Satellite 5G Internet. Available from: http://mediasat.info/2017/06/23/esa-satellite-for-5g/

[2] CEPT ECC Report "Satellite Solutions for 5G". Approved: May 18, 2018

[3] 3GPP TR 22.822. 3rd Generation Partnership Project; Technical Specification Group Services and System Aspects; Study on Using Satellite Access in 5G; Stage 1 (Release 16)

[4] 3GPP TR 38.913. 3rd Generation Partnership Project; Technical Specification Group Radio Access Network; Study on Scenarios and Requirements for Next Generation Access Technologies; (Release 15)

[5] Recommendation ITU-R M.2083. IMT Vision-"Framework and overall objectives of the future development of IMT for 2020 and beyond". Accessed: September 2015

[6] 3GPP TR 22.737. 3rd Generation Partnership Project; Technical Specification Group Services and System Aspects; Study on architecture aspects for using satellite access in 5G (Release 16)

[7] Khan F. Mobile Internet from the Heavens. Richardson, Texas, USA: Samsung Electronics; 2015

[8] Eneberg J. Satellite Role in 5G, Inmarsat; 2017

[9] 3GPP TR 38.104. 3rd Generation Partnership Project; Technical Specification Group Radio Access Network; NG-RAN; Architecture description (Release 15)

[10] 3GPP TR 38.821. 3rd Generation Partnership Project; Technical Specification Group Radio Access Network; Solutions for NR to support non-terrestrial networks (NTN) (Release 16)

[11] Boeing Seeks Permission to Launch Satellite Constellation in Same V-Band Spectrum as 5G Systems. Available from: https://www.fiercewireless.com/tech/boeing-seeks-permission-to-launch-satellite-constellation-same-v-band-spectrum-as-5g-systems/

[12] Boeing Company. Application for Authority to Launch and Operate a Non-Geostationary Low Earth Orbit Satellite Systemin the Fixed Satellite Service. FCC, [Accessed: June 22, 2016]

[13] SaT5G Project Demonstrates 5G over Satellite and Holds Industry Briefing at University of Surrey. Available from: https://www.sat5g-project.eu/sat5g-industry-day-27-november

Chapter 6

An LTE-Direct-Based Communication System for Safety Services in Vehicular Networks

Shashank Kumar Gupta, Jamil Yusuf Khan and Duy Trong Ngo

Abstract

With the expected introduction of fully autonomous vehicles, the long-term evolution (LTE)-based vehicle-to-everything (V2X) networking approach is gaining a lot of industry attention, to develop new strategies to enhance safety and telematics features. The vehicular and wireless industries are currently considering the development of an LTE-based system, which may co-exist, with the IEEE 802.11p-based systems for some time. In light of the above fact, our objective is to investigate the development of LTE Proximity Service (ProSe)-based V2X architecture for time-critical vehicular safety applications in an efficient and cost-effective manner. In this chapter, we present a new cluster-based LTE sidelink-based vehicle-to-vehicle (V2V) multicast/broadcast architecture to satisfy the latency and reliability requirements of V2V safety applications. Our proposed architecture combines a new ProSe discovery mechanism for sidelink peer discovery and a cluster-based round-robin scheduling technique to distribute the sidelink radio resources among the cluster members. Utilizing an OMNET++ based simulation model, the performance of the proposed network architecture is examined. Results of the simulation show that the proposed algorithms diminish the end-to-end delay and overhead signaling as well as improve the data packet delivery ratio (DPDR) compared with the existing 3GPP ProSe vehicle safety application technique.

Keywords: clustering, D2D, LTE, proximity services, resource allocation, safety applications, vehicular ad hoc network, V2V, V2X

1. Introduction

A vehicular communication system is one of the key components of intelligent transportation and traffic management systems. Advanced traffic management systems are expected to improve traffic flow, reduce congestions and accidents, and optimize the energy consumption of vehicles. Vehicular communication systems should enable just in time data exchange mechanisms among different elements of traffic management. Early versions of the vehicular networks were developed primarily to support V2V communications which are now evolving to vehicle-to-everything (V2X) communications mode [1]. A V2V system enables vehicles to exchange messages within the close vicinity of a Host Vehicle (HV), whereas the V2X service enables the vehicle to exchange information among any data devices in the vehicular network or in the infrastructure network. The enhanced features of vehicular networks are increasing the need for more flexible communication network architecture that can support diversified services, from time-critical safety services to high data rate entertainment services. The time-critical safety services are key features of the vehicular networks to reduce traffic accidents and offer better road safety services. Hence the role of the communication network will be crucial in a vehicular network.

The vehicular ad hoc network (VANET) architecture was initially developed using the dedicated short-range communication (DSRC) and the IEEE 802.11p networking standards [2]. The main objective of the VANET is to support V2V and vehicle-to-infrastructure (V2I) communication modes. The IEEE 802.11p network uses the random-access medium access control protocol carrier-sense multiple access with collision avoidance (CSMA/CA) to support V2V and V2I services. The advantages of the CSMA/CA protocol are in its simplicity, minimum control signaling, and the broadcast nature of transmission. These enable low packet transmission delay at lower teletraffic load. However, due to the lack of coordination among transmitters, packet collisions can occur which can increase the packet transmission delay as well as reduce the packet delivery ratio. Also, the performance of an IEEE 802.11p network is affected by the network node densities which could vary on roads depending on the road layout, congestions, and time of the day. Hence the main bottlenecks of an IEEE 802.11p vehicular network are the scalability and lack of adequate Quality of Service (QoS) support for a different class of services. However, the IEEE 802.11p standard-based vehicular network technology has matured, and many commercial products are now available [3, 4]. With the introduction of 5G technologies, the transportation and ICT industries have refocused their attention to developing new systems and products mainly relying on the Long Term Evolution (LTE)-based technologies [5].

The LTE standard is commonly used as the 4G broadband wireless technology which is further evolving as one of the major components of the 5G technology [6]. The LTE is a wide-area wireless networking technology standard that uses the conventional cellular network architecture and uses direct radio communication between the user equipment (UE) and the base station commonly known as the eNodeB (eNB) as shown in **Figure 1**. The Enhanced UMTS Terrestrial Radio Access Network (E-UTRAN) represents the radio access network where the eNB and user equipment (UE) are located. The Evolved Packet Core Network (EPC) connects the radio access networks and the external network such as the Internet. The core network hosts various control entities, databases, and functional servers. Cellular networks have several benefits such as wide-area coverage, high data rate, and guaranteed QoS for multiple services. However, the conventional centralized cellular networks are not always suitable for vehicular networks to support some of the services particularly for distributing time-sensitive broadcast services such as the Cooperative Awareness Message (CAM). In a conventional cellular network, all data communication between devices must go through the eNB, irrespective of whether they are located next to each other or at a long distance. The CAMs are transmitted from each vehicle to its neighboring vehicles to distribute situational awareness information.

The CAMs are periodic messages that have a 10 Hz generation frequency with latency restrictions of 100 ms. In the 802.11p-based VANET, the CAM messages are broadcasted to the neighboring vehicles using the CSMA/CA protocol. Generally, conventional cellular networks can support unicast, broadcast, and multicast communications; however, these configurations are not suitable for the CAM message transmissions due to high signaling overhead. To accommodate the needs of vehicular networks, the 3GPP has started to standardize the LTE-V standard to

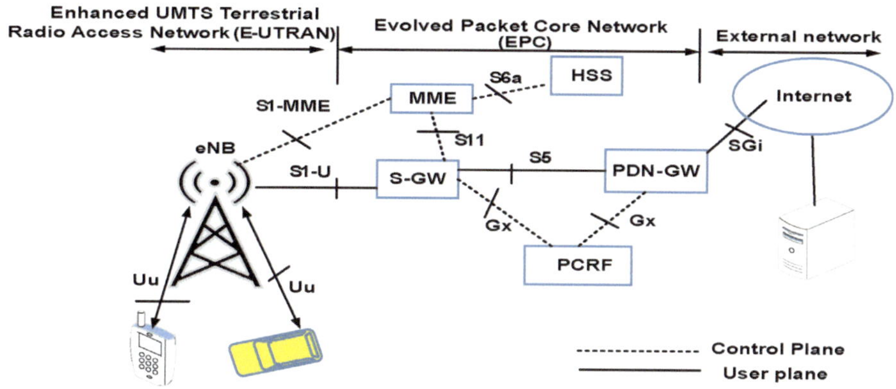

Figure 1.
LTE network architecture.

support V2X services which encompass three modes of communications: V2V, V2I, and vehicle-to-pedestrian (V2P) in Release 14. To support vehicular networking requirements, the standard has developed a new channel architecture using the PC5 interface. The standard also supports the conventional Uu interface for different vehicular services. The PC5 interface includes the sidelink which has D2D communication abilities developed under Release 12 of the LTE standard. Release 12 was mainly developed for public safety applications. The V2X communication services are being enhanced in the LTE Release 15 and will be further enhanced in Release 16.

In this chapter, we firstly review the vehicular networking and service requirements. Following the review of networking and service requirements, we briefly review the LTE-V/LTE-V2X standard. The discussion then focuses on our new algorithm referred to as Cluster-Based Cellular Vehicle-to-Vehicle (CBC-V2V) combined with a new peer discovery model referred to as Evolved Packet Core Level Sidelink Peer Discovery (ESPD). The chapter also presents the performance analysis of the CBC-V2V algorithm and compares the performance of the algorithm with other standard algorithms. In Section 2, we present the review on future vehicular network requirements. In Section 3, we briefly introduce the LTE-V/VX standard. In Section 4, our proposed LTE standard-based vehicular network resource allocation algorithm is presented. In Section 5, we present the simulation model developed to analyze the performance of the CBC-V2V algorithm. Conclusions are drawn in Section 6.

2. Future vehicular network requirements

Traffic management systems are constantly evolving to improve road traffic services and the safety of road users. Recently, the 3GPP introduced a number of vehicular network use cases in the LTE-V2V Release 14 [7] for future vehicular networks. The study showed that the vehicular network requirements have evolved over time. In early days, vehicular networks were developed mainly to support safer vehicle movements and reduce traffic congestion. However, future vehicular networks are planning to support a range of basic and enhanced services. Some of the future suggested services are listed below. The following list shows that future vehicular network requirements have been extended to include several smart city services such as parking management services, pedestrian and vulnerable road user safety. These services need to be supported by four different network configurations, i.e., V2V, V2I, V2P, and Vehicle-to-Network (V2N). Some of the service characteristics are briefly summarized in **Table 1**.

- Forward collision warning (FCW)
- Control loss warning (CLW)
- Emergency vehicle warning
- V2V emergency stop
- Cooperative Adaptive Cruise Control (CACC)
- V2I emergency stop case
- Queue warning
- Road safety services
- Automated parking system (APS)
- Wrong-way driving warning (WDW)
- V2X message transfer
- Pre-crash sensing warning
- V2X services in areas outside network coverage
- V2X road safety services via infrastructure
- V2N traffic flow optimization
- Curve speed warning
- Warning to pedestrian messaging
- Vulnerable road user (VRU) safety

Table 1 shows that communication needs and service requirements of future vehicular networks are quite diverse with variable QoS requirements. It is expected that over time, the service categories will grow, and their requirements will evolve. To support the above multiservice requirements, the current IEEE 802.11p networks will not be adequate due to higher traffic volume and inadequate QoS support for multiservice networks. Also, some of the services such as emergency vehicle warning or curve speed warning may need longer transmission ranges and may also increase the collision probability in CSMA/CA-based IEEE 802.11p networks. Another important consideration for the future vehicular network is the support of autonomous vehicles that require low delay and low loss reliable communication networks. Hence, the main objective of the LTE-V/LTE-V2X standard is developing an advanced cellular-based vehicular network. In the following section, we review the LTE-V2X standard based on Release 14.

Service	Main purpose	Communication mode	Service requirements
Forward collision warning	The FCW service has been proposed to warn the driver of a host vehicle (HV) about an impending rear end collision with a remote vehicle (RV) or vehicles. The FCW service can help reduce collisions	HV and RV communicate using V2V transmission mode	Periodic broadcast CAM message, support high mobility, early warning message
Control loss warning	The CLW service enables an HV to broadcast self-generated loss of control message to RVs. Upon receiving the message, RVs warn drivers for appropriate action(s)	HV and RV communication using V2V services	Communicate messages over a distance to generate warning message with ample time to respond. Event-based broadcast message
Emergency vehicle warning	This service enables all vehicles to acquire location, speed, and direction information of surrounding emergency vehicle(s) to assist smooth movement of emergency vehicles	V2V communication using LTE-D2D	Event-based CAM message broadcast to cars within 300–500 meters
Cooperative Adaptive Cruise Control (CACC)	The CACC service provides convenience and safety benefits to group of vehicles in close vicinity. Can be used for platooning structure	Mainly V2V services, but V2X communication can also be used to obtain forward traffic flow information	The service can support a maximum latency of 1 sec and a maximum frequency of one message per second
Queue warning	This service allows vehicles to receive forward road queue warning messages. Road user safety can be significantly increased by using this service	V2V and V2I communication services	Able to transmit and receive V2I messages with a maximum relative velocity of 160 km/h. Support an appropriate communication range necessary for early warning
Road safety services	Using this service, V2X messages are delivered from an UE to other UEs via an installed Road Side Unit.	V2X and V2I services	A V2X message should be delivered within 100 ms via an RSU with low delivery loss. An RSU should be able to transmit V2X messages at a maximum frequency of 10 Hz
Curve speed warning	This application sends alert messages to the driver to manage possible blind spot or the curve at an appropriate speed. An RSU is placed before a curve to transmit information such as curve location, recommended speed, curvature, and road surface conditions	RSU-based I2V and V2I services	I2V message transmission with a maximum latency of 1 sec and maximum frequency of one message per second

Table 1.
Service characteristics.

3. LTE-V2X standard

The LTE standard is widely used in public and private mobile radio networks. LTE technology has been identified to support vehicular network services using V2X architecture. The V2X service architecture is shown in **Figure 2**. As mentioned in the previous section, the V2X communication services include four different modes of communication (V2V, V2I, V2P, and V2N). These links are bidirectional. 3GPP study groups in collaboration with transport industries have started standardization activities on LTE-based vehicular networks in the working group 1. After several studies and developing several initial specifications on V2X services based on LTE, Release 14 was published in 2017 [8]. The standard is further developed in Release 15 in 2018 supporting enhanced V2X networking features. The enhancements go beyond the support of CAM and Decentralized Environmental Notification Messages (DENM) transmissions as shown in **Table 1**. The 3GPP specifications did not allocate any specific frequency band to support V2X services. European Telecommunications Standard Institute (ETSI) has allocated a 70 MHz spectrum in the 5.9 GHz band in which there is no overlap between V2X and conventional cellular network services. This separation of operating frequency will enable different operators to provide vehicular network services independent of conventional mobile operators. The 5.9 GHz LTE band will allow the system to coexist with IEEE 802.11p-based systems. However, the mobile operators can also use the licensed band to support the V2X services. The V2X services can use the conventional air interface as well as the newly developed D2D interface using the sidelink channel. The D2D communication architecture is briefly introduced in the following section.

3.1 D2D communication architecture

The LTE-V2X architecture has been developed to support diverse vehicular network services as discussed above. The architecture uses the new air interface PC5 along with the conventional Uu interface to support various services. The PC5 interface can offer enhanced network services such as device-to-device communication, normally supported by the ad hoc network architecture. The device-to-device communication services was introduced in Release 12 which was originally developed for the safety services [9]. The LTE Release 12 architecture is shown in **Figure 3**. The figure shows a new service function the Proximity Service located in the Evolved Packet Core which allows the devices to discover peer devices for D2D

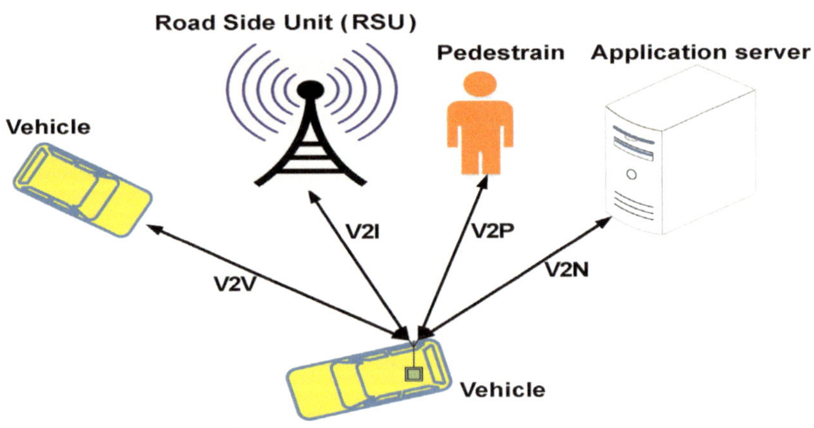

Figure 2.
V2X communication architecture.

communication services. The ProSe function allows users to directly communicate and exchange data with neighboring devices by sending a registration message to the eNB with a ProSe application ID. The eNB organizes the communication between the devices using the control channels. Once the communicating devices are matched by the eNB, then they can directly communicate using the PC5 interface as shown in **Figure 3**. The PC interface functions are summarized in **Table 2**. Details of these interfaces can be found in [10].

The channels in the Uu and PC5 interfaces are organized as logical, transport, and physical channels. **Figure 4** shows the mapping structure of these channels used for the sidelink communication in the LTE standard. There are two logical channels introduced for sidelink communication: first is the SL Traffic Channel (STCH), and second is SL Broadcast Control Channel (SBCCH). The STCH is an interface to the Physical SL shared Channel (PSSCH), which transports the data carrying user information over the air. The SBCCH is used to broadcast control data, for synchronization in the out of coverage or partial coverage, or for the synchronization between UEs which are located in different cells. There is also a Transport and Physical Sidelink Control Channel carrying the SL control information (SCI). There is a new transport and physical channel for direct discovery: sidelink discovery channel (SL-DCH) and the physical sidelink discovery channel (PSDCH).

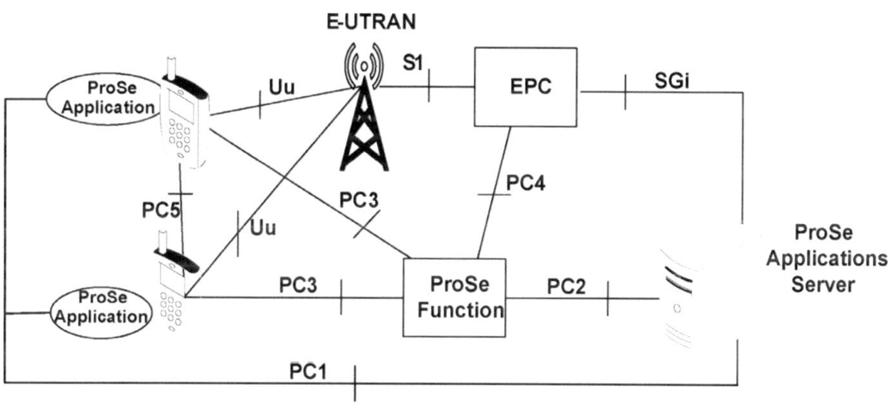

Figure 3.
LTE release 12 D2D reference network architecture [9].

Interface	Main functions
PC1	The ProSe application server can communicate towards a ProSe application in the UE through the interface
PC2	The ProSe application server can communicate with the ProSe function through this interface
PC3	The ProSe function can connect to the UE through the PC3 interface
PC4	The ProSe function connects with Evolved Packet Core in the network through PC4 interface
PC5	A PC5 interface enables direct communication between two UEs

Table 2.
PC interfaces.

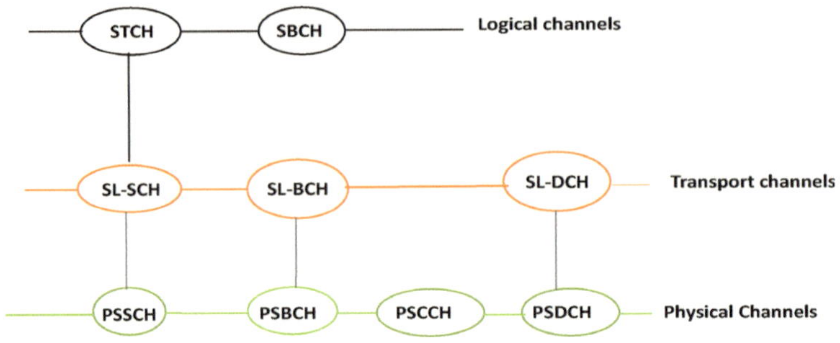

Figure 4.
Mapping of channels for sidelink communication in 3GPP LTE.

3.2 Enhanced D2D communication architecture for V2X communications

Recently, several fundamental modifications have been carried out to enhance the PC 5 interface in the Release 14 to support V2X operational scenarios and requirements as shown in **Table 1** [11]. The sidelink LTE-V2X employs the single-carrier frequency division multiple access (SC-FDMA) which permits the UE to access radio resources in both time and frequency domains. In the frequency domain, the subcarrier spacing is fixed to 15 kHz, and subcarriers are utilized in groups of 12 (i.e., 180 kHz). To support different V2X operational requirements, the transmission channels may use a higher carrier frequency of 6 GHz with very high relative velocity. However, due to the high relative velocity and the use of higher carrier frequency, inter-carrier interference (ICI) due to higher Doppler shift and insufficient channel estimation due to shorter coherence time could be a problem compared to the legacy 3GPP systems.

To improve the performance in the presence of high Doppler shift, the sidelink interface has been tuned to counteract the severe Doppler shift experienced at high speed. In the time domain, additional demodulation reference signal (DMRS) symbols have been added in one subframe to handle the high Doppler shift associated with relative speeds of up to 500 km/h and the use of higher carrier frequency [12]. The new subframe structure is illustrated in **Figure 5**. Fourteen symbols form a subframe of 1 ms, also called transmission time interval (TTI), which include nine data symbols, four demodulation reference signal (DMRS) symbols, and one empty symbol for Tx-Rx switch and timing adjustment. The LTE-V2X has a large number of modulation and coding schemes (MCS), with 4-QAM and 16-QAM modulations, and an almost continuous coding rate. The minimum radio resource allocated to an

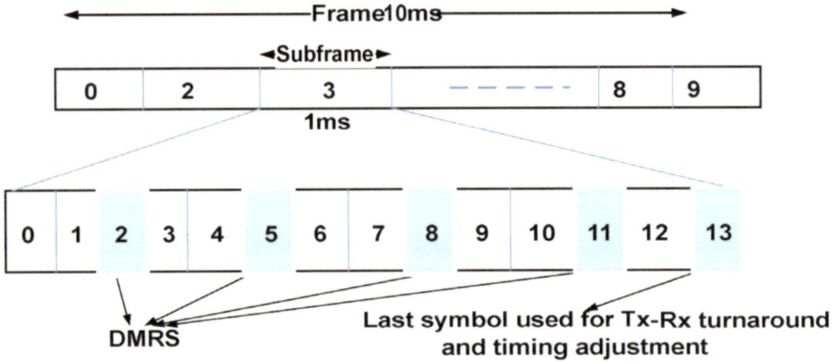

Figure 5.
V2V subframe for PC-5 interface structure [12].

LTE-V2X link is the subchannel in the frequency domain, corresponding to a multiple of the 12 subcarriers groups, and the TTI in the time domain. One packet normally occupies one or more subchannels in a TTI. To improve the system-level performance under high node density while meeting the latency requirement of a V2V link, a new classification of scheduling assignment and data resources is designed where the scheduling assignment is transmitted in sub-channel using specific Resource Blocks (RBs) across the time. More specifically, each data packet also known as Transport Block (TB) has an associated control message called the Sidelink control information (SCI). TB and the associated SCI must be transmitted in the same subframe but can be allocated in adjacent and nonadjacent resource blocks.

Figure 6 depicts the overall network architecture enhancement in Release 16 for V2X services [13]. Two new entities are introduced: the V2X Application server and the V2X control function to support the V2X services. The V2X control function is the logical function that is used for network-related actions required for V2X. The parameters required for V2X communications can be obtained from V2X Application Server. It is also provision the UEs with Public Land Mobile Network (PLMN) specific parameters that allow the UE to use V2X in this specific PLMN. The V2X Application server incorporates the V2X capability for building the application functionality. It is responsible for receiving uplink data from the UE in the unicast mode, providing the parameters for V2X communications over the PC5 reference point to V2X control function. As per the network architecture, several new reference points (or interface) have been introduced. The roles of V2X reference points are summarized in **Table 3**.

To support the V2X communication, Release 14 introduced the new communication modes (mode 3 and mode 4) as shown in **Figure 7**. Mode 1 from Release 12 was enhanced to mode 3 for V2X communication; similarly, mode 2 from D2D was enhanced to mode 4 for V2X. In mode 3, the UEs' resource reservation and scheduling are performed by the eNB, while in mode 4 the UEs choose the radio resources autonomously. Mode 3 algorithms are not defined in the specifications and their implementation is left to vendors. In contrast, mode 4 can operate without cellular coverage and is therefore considered as the baseline V2V mode since safety applications cannot always depend on the availability of cellular coverage. In mode 4, also known as autonomous or out-of-coverage, each node selects the resources based on a sensing procedure and a semi-persistent scheduling (SPS) mechanism. Mode 4 includes a distributed scheduling scheme for vehicles to select their radio resources and includes the support for distributed congestion control. The detailed

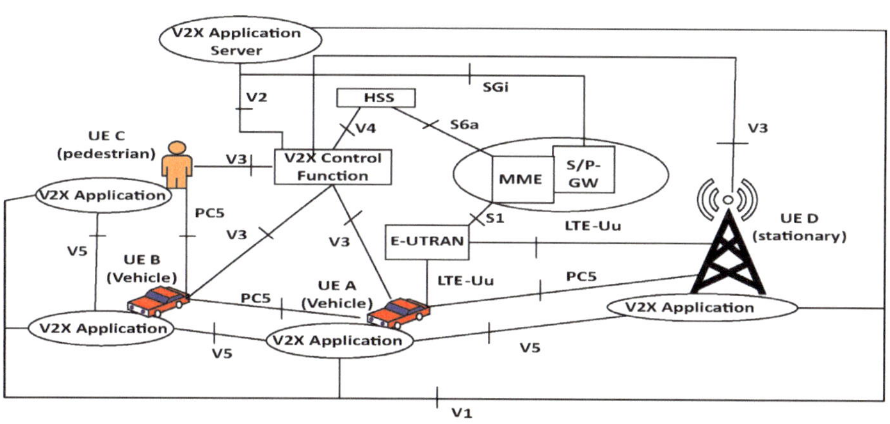

Figure 6.
Enhanced ProSe D2D sidelink architecture for V2X communications [13].

Interface	Main functions
V1	The V2X application server can communicate towards an V2X application in the UE through V1 interface
V2	The V2X application server can communicate with the V2X control function through V2 interface. The V2X application server may connect to V2X control function belonging to multiple PLMNs
V3	The V2X control function can connect to the UE through the V3 interface
V4	The V2X control function connects with entity Home Subscriber Server (HSS) in Evolved Packet Core in the 3GPP network through V4 interface
V5	A V2X application in UE can communicate towards a V2X application in different UEs through V5 interface
SGi	An EPC can connect to the V2X application server through SGi interface

Table 3.
V2X interfaces.

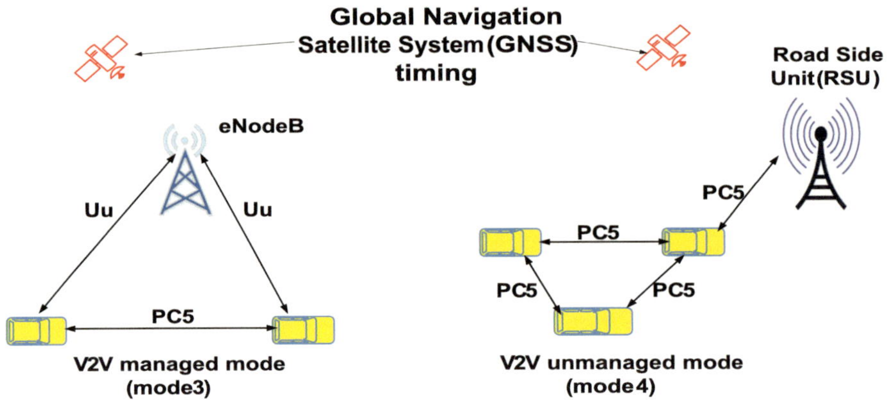

Figure 7.
V2X communication mode defined in release 14.

description by 3GPP for mode 4 algorithm is presented in [14, 15]. The Global Navigation Satellite System (GNSS) is introduced to provide accurate timing and frequency references in the off-coverage scenario [16].

3.3 Review on current research on LTE vehicular networks

Since the LTE Release 14 was standardized, several studies have been carried out to compare the performance of IEEE 802.11p and LTE-V2X vehicular networks. In [17], comparative experiments with real devices were carried out, demonstrating improvement of the C-V2X system performance. The work demonstrated that the latency in C-V2X under congested conditions can be maintained under 100 ms.

The use of cellular technologies for vehicular networks has been investigated to meet the requirements of safety services in [5, 18, 19]. The work showed that traffic hazard warning messages are disseminated in less than a second. Hybrid architectures based on the LTE and the 802.11p standards have been proposed to exploit the benefits of both networks [20, 21]. Sivaraj et al. [20] present a cluster-based centralized vehicular network architecture which uses both the 802.11p and the LTE standards for well-known urban sensing application and floating car data (FCD) application. The authors also compared those system performances with other decentralized clustering protocols. Remy et al. [21] propose a cluster-based VANET-LTE hybrid architecture for multimedia-communication services.

In [22], the authors provide the delay performance analysis of hybrid architectures. Calabuig et al. [22] propose a hybrid architecture known as the VMaSC-LTE that integrates the LTE network with the IEEE 802.11p-based VANET network. In [22], the authors propose a Hybrid Cellular-VANET Configuration (HCVC) to distribute road hazard warning (RHW) messages to distant vehicles. In this hybrid architecture, cluster members (CMs) communicate with the cluster head (CH) by using the IEEE 802.11p link, and the CHs communicate with the eNB by using cellular links. However, this proposed 802.11p-LTE hybrid architecture increases the transmission delay at the same time as reducing the reliability when the IEEE 802.11p-based network needs to support higher node densities, leading to higher medium access delays. Toukabri et al. [23] propose a Cellular Vehicular Network (CVN) solution as a reliable and scalable operator-assisted opportunistic architecture that supports hyper-local ITS services for the 3GPP Proximity Services. A hybrid clustering approach is suggested to form a dynamic and flexible cluster managed locally by the ProSe-CHs. However, the authors do not focus on the transmission of safety messages in the network.

In [24–26], the authors compare the performance of the IEEE 802.11p and the LTE-V2X in terms of reliability. They mainly used simulation with a moving vehicle and consider the highway scenario to analyze the performance of two technologies. Some of them also include an urban Manhattan case [25, 26]. Bazzi et al. [27] compare IEEE 802.11p and LTE-V2V for cooperative awareness in terms of maximum awareness range and also provides analytical evaluation of the proposed schemes. Min et al. [25] introduce a resource scheduling algorithm known as Maximum Reuse Distance (MRD) for V2V communication under network coverage. The proposed scheduling algorithm is in-line with Cellular-V2X mode 3 with the aim of minimizing the interference and increasing the reliability and latency of V2V communication.

Recently, a global alliance called the Fifth Generation Automotive Association (5GAA) has developed a model to assess the relative performance of LTE-V2X (PC5) and the IEEE 802.11p technologies with regard to improving the safety, focusing on direct communications [28]. This study indicates that the LTE V2X (PC5) outperforms the 802.11p in reducing fatalities and serious injuries on European roads. All of the abovementioned works agree that LTE-V2X can provide better performance compare to IEEE 802.11p. This is due to a combination of the superior performance of LTE-V2X (PC5) at the radio link level for ad hoc/direct communications between road users. However, the use of LTE-V2X for vehicular applications is not mature yet. In particular, LTE-V2V devices are still under development, and the allocation (and management) of radio resources is still under investigation.

4. CBC-V2V system model

In this section, we present an LTE-based cellular network architecture for V2X communication using the PC5 interface of the LTE standard. We assume that all vehicles on the road are within the coverage of the eNB. A highway road traffic scenario is considered where traffic is flowing in both directions in a multilane road as depicted in **Figure 8**. We assume that each vehicle is equipped with a GPS device capable of providing accurate position measurements. The highway is partitioned into fixed-size regions known as a cluster. Vehicles on the road with near

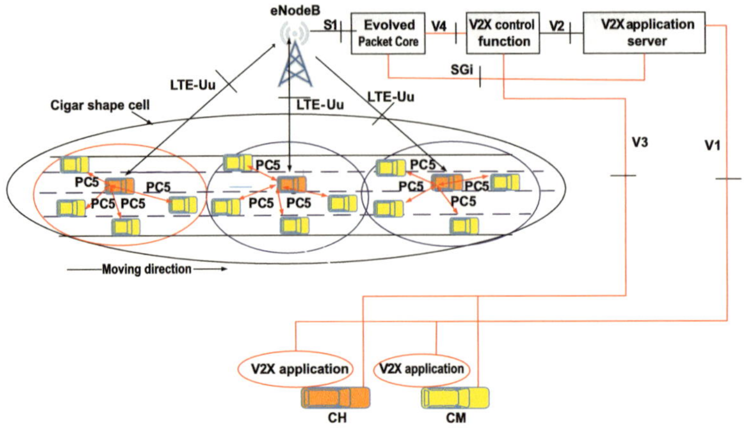

Figure 8.
Highway scenario for proposed cluster-based V2V cellular (CBC-V2V) architecture.

proximities form a cluster where they exchange the safety messages to each other using a CBC-V2V-based packet transmission technique.

We are considering two types of vehicles: the first type represents the user terminals capable of acting as a CH and supports D2D communication using the PC5 interface. The CH also manages the network resource usage among the group of devices communicating over D2D links. The second type of vehicles represents the network devices that can only act as CMs. These vehicles connect to the appropriate CH to assist them in establishing the D2D links to exchange messages. In this model, a vehicle uses two communication links: the conventional Uu channels and the D2D links using the PC5 interface. Cluster members can communicate with others using the PC5 links, whereas a CH communicates with the eNB using the Uu interface. Although the D2D channels enable two neighboring UEs to communicate directly, all signaling and data transmission processes should still be under the control of the eNB in order to comply with the LTE-Advanced architecture requirements.

4.1 Cluster-based cellular V2V (CBC-V2V) communication architecture

We propose a cluster-based cellular V2V communication architecture that combines the new sidelink peer discovery model to support safety services. We propose to use a cluster topology where communication among cluster members is coordinated by the cluster shown in **Figure 8**. Vehicular networks are generally dynamic where vehicles may arrive new in a cluster location or may leave a cluster. For a newly arrived vehicle, it is necessary to find out necessary system information to join an appropriate cluster. In the following section, our proposed sidelink peer discovery model is presented. Following that discussion, our cluster-based cellular V2V communication mechanism combining with a round-robin scheduling technique is proposed to distribute the radio resources among the cluster nodes.

4.2 EPC level sidelink peer discovery (ESPD) model

For direct communication, two devices must be aware of each other. ProSe peer discovery is the first step to start a direct transmission. Since the introduction of D2D communication architecture in Release 12, many device/peer discovery techniques have been developed using two models defined in the standard. From the user's perspective, they can be classified into restricted discovery and open discovery [29]. For restricted discovery, the user entity is not allowed to be detected without its explicit permission. In this case, it prevents other users to distribute their information to protect user privacy. It suits social network applications (e.g., group gaming and context sharing with friends). For open discovery, a user entity can be detected as long as it is within another device's proximity. From the network's perspective, device discovery can be divided into two types: direct discovery and Evolved Packet Core (EPC) discovery. UE would search for a nearby device autonomously; this requires a UE device to participate in the device discovery process. Direct discovery work in both in-coverage and out-of-coverage scenarios. There are also provisions for EPC level discovery that notifies the terminal about other users detected in the vicinity based on the user interest information and the UE location information registered by terminals in the ProSe function [30].

All vehicles that need to use the D2D link must have the ProSe capability features: the ability to discover, to be discovered, and to communicate with discovered devices. Within the existing EPC level discovery model, the ProSe function authenticates the user by checking its credential with the HSS as to whether the user is permitted to utilize ProSe features. After successful authentication of the UE, the ProSe function creates an EPC ProSe Subscriber ID (EPUID) and assigned it to the registered device. Once a vehicle registered as a ProSe subscriber, it can run the applications that support proximity services, named as a ProSe-enabled applications. The application server allocates the user an Application Layer User ID (ALUID) to recognize him within the context of this particular application.

However, these device discovery and the EPC level discovery models require significant control signaling or message exchanges such as announce requests, monitor requests, match reports, etc. [30, 31]. Our proposed discovery mechanism diminishes network resource requirements. It assumes that every vehicle is equipped with a GPS receiver and can accurately determine its position and direction of movement. **Figure 9** appears the signaling diagram of the proposed EPC level discovery technique elaborated as follows:

1. When a new vehicle reaches an eNB coverage area, the downlink frame synchronization is accomplished once it has decoded the primary synchronization signal (PSS) and the secondary synchronization signal (SSS) messages, which are accessible on the downlink broadcast control channel.

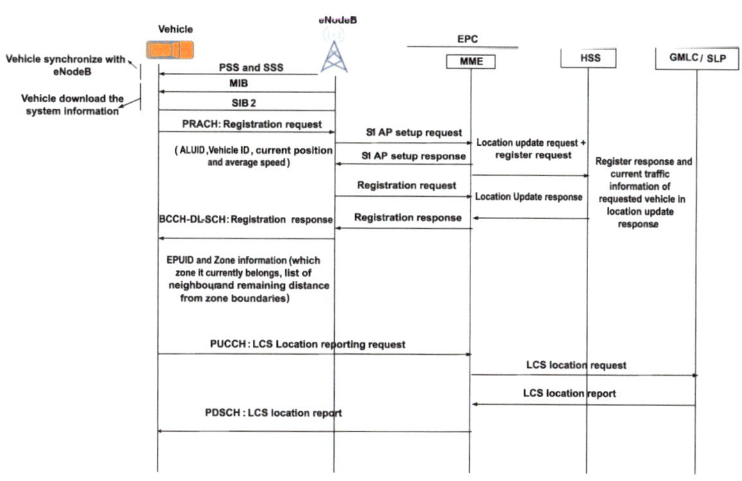

Figure 9.
EPC level Sidelink peer discovery (ESPD) model for VANET.

The vehicle at that point downloads the Master Information Block (MIB) from the broadcast channel. This channel incorporates the downlink and uplink carrier configuration information. Further, the vehicle utilizes the Downlink Shared Channel (DL-SCH) to download the system information block. The SIB2 block contains necessary parameters for the initial access transmission.

2. In the initial state, each vehicle on the road must register itself with the eNodeB using its current GPS position. Unlike the existing EPC level discovery, the vehicle sends its location information in the registration request to the eNodeB instead of using the ProSe function for user (vehicle) registration. The vehicles will forward their information (such as ALU_ID, current GPS location, an average speed of the vehicle, discovery range, and vehicle ID) in the registration request message utilizing the Random Access Channel (RACH) to the eNB. The eNB acknowledges the registration request and broadcasts the registration response back to vehicles along with the current traffic profile over the broadcast channel. The vehicle's traffic profile contains an EPC ProSe Subscriber ID, zone information (i.e., to which zone it currently belongs), neighboring vehicle list, and the vehicle's remaining distance from its location.

3. After accepting the information supplied in the registration response, the vehicle collects all the data in its Vehicle Information Register (VIR), a repository that stores vehicle and surrounding information. For D2D communication, each vehicle updates its neighborhood table with a new list of neighboring vehicles and builds knowledge of its local environment. The global mobile location center (GMLC) keeps vehicle locations tracked. Once the vehicle comes to a new zone or crosses the boundary of the zone, the location alert, i.e., the Location Service (LCS) report, will be received and vehicle will require re-registration to update its VIR.

4.3 Cluster formation

After the peer discovery, each vehicle needs to select an appropriate Cluster Head (CH) to associate with it. Using the peer discovery model, after successful registration, each vehicle updates its Neighborhood Table (NVT) in its VIR with the new proximity data (i.e., a list of neighbor vehicles) along with the vehicle ID, total number of vehicles, and current state of the each vehicle in the list. Once the new proximity data received, the vehicle will reach in the Selection State (SE). As shown in **Figure 10**, a vehicle in the selection state first tries to connect to the existing cluster to minimize the number of clusters. Hence, the source vehicle (SV) first checks the total number of vehicles, their position conjointly, and the state of each vehicle in its NVT.

If the vehicle finds a cluster head in its NVT, and the number of members in the cluster is lower than the maximum number of members allowed, the SV will attempt to connect to the existing CH. In the NVT, if none of the neighboring vehicles are listed as CH or the vehicle is unable to connect to any of the neighboring CHs, the vehicle inspects the neighboring vehicles in the semi-cluster head (SCH) state. If there are vehicles in the SCH state in its NVL, the source vehicle tries to connect the existing semi-cluster head. If none of the neighbor vehicles are listed as CH or SCH, the SV checks the neighboring vehicle in Selection State. If the SV discovers the vehicles in SE in NVT and it has the lowest average speed and the maximum distance from its current location to the zone boundary (i.e., longest lifetime) among them, then it will take the role of CH. Otherwise, the SV becomes

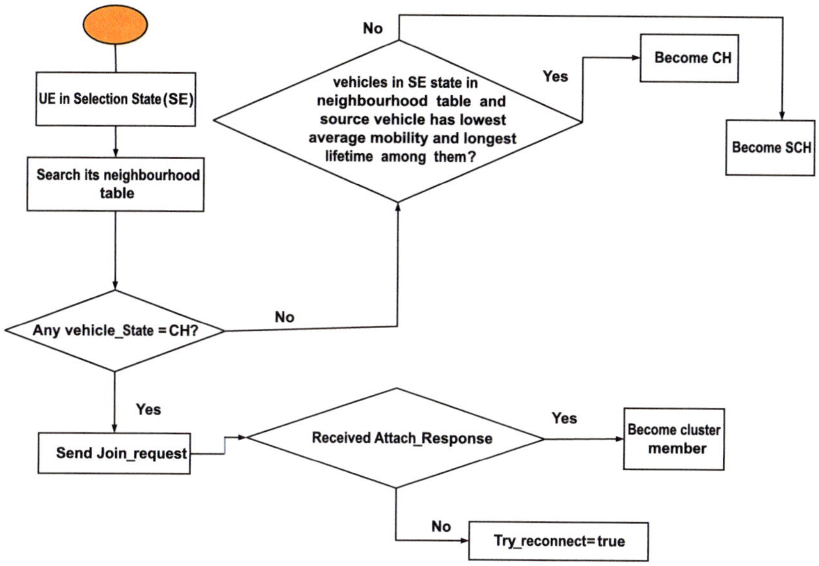

Figure 10.
CBC-V2V clustering approach.

an SCH. SCH is the state the vehicle has no potential neighboring vehicle that can connect to it.

4.4 Cluster head and semi-cluster head selection

Upon receiving the new proximity data in a neighboring table, an SV search the NVT during the time period T_{search} to check the vehicles in CH, SCH, and SE state. If none of the neighbor vehicles are recorded either as CH or SCH, the vehicle will check the neighboring vehicles in the SE states. If there are the vehicles in SE state in the NVT and the SV has the most reduced average speed and a maximum distance from its current location to the zone boundary (i.e., longest lifetime), at that point it becomes the CH. The algorithm for the CH and the SCH selection is presented in Algorithm 1. Each vehicle calculates its average speed periodically. If none of the neighbor vehicles are recorded either as CH, SCH, or SE, a source vehicle will take the role of SCH. In case the vehicle in the SCH state gets any joining request from a neighboring vehicle during the time period T_{SCH}, then it will take the role of the CH. Otherwise, it will reach in the SE state and require a re-registration to receive new proximity data.

Algorithm 1. CH and SCH selection

---

while $T_{search} \neq 0$ && there is no potential neighbouring to connect ($V_{State} \neq CHorSCH$) do
  if $V_{State} = ALL_{SE}$ then
    The SV will compare its S_{SV} and T_{Life} with other vehicles in NVL;
  if $S_{SV} < S_{ALL}$ and $T_{Life} > T_{ALL}$ then
    $SV \rightarrow CH$;
    else
    $SV \rightarrow SCH$;
    end if
  end if

---

if $T_{SCH} \neq 0$ then
  SCH_i receive any joining request from neighbouring vehicle;
  $SCH \rightarrow CH$;
else
  $SCH \rightarrow SE$;
end if
end while

4.5 V2X sidelink channel structure

Using communication mode 3, we suggest the 3GPP standard-based V2V sidelink channel structure as shown in **Figure 11**. The figure shows that an eNB reserves 10 D2D subframes on uplink cellular traffic channels in the time division multiplex (TDM) manner. The D2D subframe repetition rate is 100 ms. Each subframe contains two slots; hence a single carrier offers 20 slots for sidelink communications. The RBs are used to transmit data and control information. The data is transmitted using transport blocks (TBs) over the Physical Sidelink Shared Channels. Sidelink control information messages are transmitted over the Physical Sidelink Control Channels (PSCCH) [16]. The number of RBs in a slot depends on the bandwidth of an LTE-V network cell. Using a 3 MHz transmission bandwidth, there will be 15 RBs in 1 slot available for the D2D communication.

4.6 CBC-V2V communication

Our proposed CBC-V2V communication for safety message transmission is shown in **Figure 12**. As seen, the intra-cluster communication procedure between cluster members V_{A1} and V_{A2} belongs to a cluster CH_{A0} and inter-cluster communication from V_{A1} to the vehicle V_{B1} which belongs to a neighbor cluster CH_{B0}. For the rest of the vehicles in the network the same procedure will follow. A CH acts as a ProSe gateway node for vehicle-to-infrastructure (V2I) and infrastructure-to-vehicle (I2V) communication. The CH utilizes the Physical Uplink Shared Channel (PUSCH) uplink grant allocated during the random access procedure to send the RRC connection request along with the data structure called cluster_info. In the cluster_info, each CH keeps the information such as the CH_{ID} and the number of CMs attached to it. Based on the cluster_info in the RRC connection request, an eNB dynamically allocates resources to a CH for D2D communication. At the cluster level, each cluster head further schedules the resources among its CMs using the new cluster-based round-robin scheduling as described below.

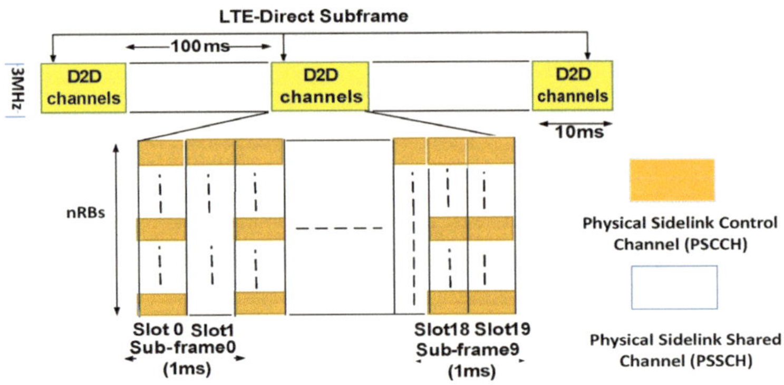

Figure 11.
V2V sidelink subframe structure.

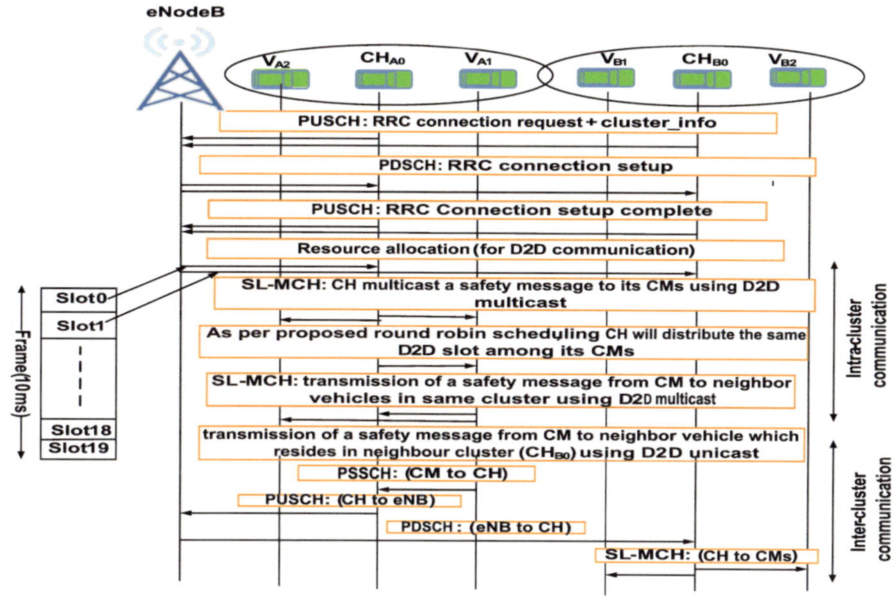

Figure 12.
CBC-V2V communication over sidelink channels.

Radio resources are initially allocated to the CH for each cluster of nodes. The CH then conducts round-robin resource scheduling among its CMs (i.e., vehicles) based on the vehicle ID. The round-robin scheduling approach is based on the idea of being fair to all active users in the long term by granting an equal number of physical resource blocks (PRBs). Our proposed resource allocation scheme is operated by dynamically assigning the same slot to the multiple users, in turn, using node IDs in ascending order. Subsequently, members of a cluster can share the same slot in turn to transmit their own CAM.

As shown in **Figure 11**, 10 subframes for D2D communication show up in every 100 ms which are shared between different clusters. Since each cluster is designated one slot, the same subframe will support two clusters. In the example, slot 1 is assigned to CH_{A0} and slot 2 is assigned to CH_{B0}. When the resource is allocated, the CH chooses PRBs within the available slot to transmit its posses CAM to its CMs in the multicast mode. A ProSe-enabled node cannot receive and decode the D2D message while it is transmitting, due to the half-duplex nature of most transceiver designs. Therefore, in the cluster, when one vehicle is transmitting, the rest of the vehicles will receive the CAM from the transmitting vehicle. Each safety message can be accommodated utilizing four PRBs based on the selected modulation and coding scheme and the packet size. On completion of the transmission from the CH, it will assign the same slot to its CMs. The next vehicle V_{A1} is thereby allocated the same slot on its turn based on its vehicle ID. Then V_{A1} multicast its own safety message to its neighboring vehicles. The same procedure will follow by the remaining vehicles in the cluster. To maximize reuse of the spectrum, the same D2D resource can be assigned to different nonoverlapping clusters.

In this architecture, the inter-cluster communication is required to share safety messages by vehicles which are found at the edge of the two neighboring clusters. In the example, vehicle V_{B1} is in the neighbor list of V_{A1} but out of range of its CH_{A0}. Therefore, direct communication is not conceivable between V_{A1} and V_{B1}. In this case, CH_{A0} collects the safety message from its cluster member V_{A1} over the D2D Physical Sidelink Shared Channel and transmits to the eNodeB over the LTE interface in the unicast mode. At that point, the eNodeB conveys the safety traffic message to a concerned neighbor CH_{B0} over the LTE interface. The CH_{B0} multicasts the safety message to its cluster members V_{B1} and V_{B2} via the LTE-D2D PC5 interface.

5. Simulation model

An OMNET++ version 5.1.1-based simulation model is developed utilizing the SimuLTE library [32] that utilizes the INET framework 3.4.0. For enhanced traffic simulation, GPS data incorporation, and mobility support, we utilized the Veins Package with a realistic mobility model generated by the microscopic road traffic simulation package: Simulation of Urban Mobility (SUMO) [33]. To add the mobility support feature in SimuLTE, a new interface known as vehicularMobility module has been added. This new mobility model can be implemented by the TraCIMobility module defined by the Veins. There is another mobility module known as INETMobility present in the INET framework. A vehicle can utilize only one mobility module during the simulation; therefore both modules (i.e., INETMobility and vehicularMobility) are defined as a conditional module within the Ned file. Veins use the OMNeT++ API to create and initialize the new module dynamically. When a new vehicle is created, it needs to obtain an IP address to communicate. SimuLTE demands the assignment of IP addresses to the IPv4NetworkConfigurator module provided by INET.

A new parameter, i.e., d2dcapable, is utilized in the .ini file to enable direct communication between two UEs. Most of the PC-5 operations at each layer of the LTE stack are created by extending pre-existing SimuLTE capacities. For each D2D competent user, an LTE binder keeps up a data structure that contains the set of directly reachable destinations. In expansion to the existing DL/UL ones in SimuLTE, a new flow path, PC-5, has been distinguished. From the UE point of view, IP datagrams reach the PDCP layer and either the PC-5 or the UL directions can be associated with the corresponding flow, depending on whether the destination is in the LTE Binder peering table or not. The detailed description of configuring D2D communication in OMNET++ with SimuLTE is given in [34]. The key simulation parameters are summarized in **Table 4**.

We modified the existing D2D communication model in the SimuLTE to support our proposed cluster-based cellular V2V architecture. **Figure 13** shows the CBC-V2V communication model consists of an access network entity (single eNodeB) and core network entities (MME, HSS, and GMLS) are utilized to support our proposed EPC Level Sidelink Peer Discovery model. In the simulation, we design a multilane highway scenario where the vehicles are distributed according to the Poisson process. The vehicles form the clusters using our proposed clustering scheme for D2D communication. To implement our proposed clustering scheme, we utilize the sample source code accessible online [35]. Each cluster node keeps up neighborhood table that contains its neighbor's ID and their state. In the simulation, we include scenarios of both multicast and unicast shown in **Figure 12**. The model is simulated for both scenarios utilizing the parameters presented in **Table 4** for 800 seconds. At the MAC layer in the SimuLTE, we modified the scheduling model (i.e., LTEDrr) to implement our proposed round-robin scheduling scheme presented in Section 4. Utilizing the proposed round-robin scheduling technique, each cluster node receives an equal share of the radio resource for D2D communication.

5.1 Performance analysis

Using the number of clusters/km and the traffic load (i.e., number of vehicles/cluster) parameters, we examine the overall end-to-end delay, resource utilization,

Parameter	Value
Maximum velocity	40–70 km/h
Number of vehicles	96 vehicles/km
Road length and number of lanes	5 km and 4 (i.e., 2 in each direction)
Carrier frequency	2.6 GHz
Duplexing mode	TDD
CAM generation rate	10 packets/sec
Transmission bandwidth	3 MHz (i.e., 15 RBs)
Path loss model	Highway scenario
Fading model	Shadowing
eNodeB Tx power	46 dBm
UE Tx Power	26 dBm (Uplink), 5 dBm (Sidelink)
Coverage range	1000 m
Noise figure	5 dB
Cable loss	2 dB
Simulation time	800 s
Packet size	340 bytes
T_{safety}	100 ms
Number of vehicle/cluster	12
$CH_{maxmember}$	11

Table 4.
Main simulation parameters.

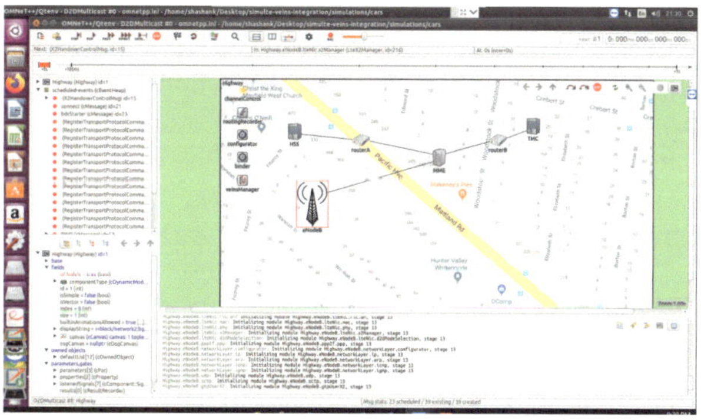

Figure 13.
CBC-V2V simulation model.

signaling overhead, and data packet delivery ratio performance of our cluster-based D2D vehicular network architecture. The following performance metrics are used to evaluate the proposed algorithm.

5.1.1 Control signaling overhead

The signaling overhead is measured for the proposed EPC level peer discovery and the D2D packet communication techniques. The overall signaling overhead of the network can be calculated as

$$X_{SO}(c) = \sum_{i \in N} (\overline{x}_{pd} + \overline{x}_{d2d}) \tag{1}$$

where $\overline{x}_{pd}$ represents the average signaling overhead in bits related to the control signaling required for the peer discovery and $\overline{x}_{d2d}$ represents the average signaling overhead related to the control signaling required for the D2D communication. $\overline{x}_{pd}$ can be calculated as the number of slots used for peer discovery out of the total number of n subframe available in the cell i as

$$\overline{x}_{pd} = \frac{x_{ir1} + x_{ir2} + x_{ir3}, \ldots, x_{irn}}{n} \times 100 \tag{2}$$

Similarly, we calculate $\overline{x}_{d2d}$ the overhead for the D2D communication and calculate the overall signaling overhead. **Figure 14** shows the signaling overhead required by the CBC-V2V, the default 3GPP ProSe algorithm, and the LTE-Advanced algorithm using conventional cellular architecture. The results clearly show that the CBC-V2V introduces lower signaling overhead compared to the other two standards which can be used in a VANET. The main reason for the performance improvement is the lower control signaling requirement for the CBC-V2V algorithm. The major benefit comes from our ESPD algorithm which requires less control message exchange for peer discovery compared to existing peer discovery models described in Section 4. Unlike the existing 3GPP peer discovery model, in the ESPD algorithm, a vehicle receives the proximity information after the successful registration which requires very less control message exchange as shown in **Figure 2**. The smaller control signaling overhead requirement will improve the performance of safety services and guarantee the timely delivery of active safety messages.

Figure 15 shows the overall resource utilization of the CBC-V2V algorithm for safety services. We compare the results with the standard ProSe solutions in terms of a number of occupied RBs. The efficient scheduler minimizes resource utilization and distribution levels. In the CBC-V2V, each of the CH acts as a scheduler and distributes the resources among its CMs using our proposed round-robin scheduling. Two clusters can be served in a single subframe, and nonoverlapping clusters can share the same resource. Resource utilisation of the CBC-V2V algorithm is lower compared to 3GPP ProSe algorithm due to lower control signal requirements, cluster architecture and efficient resource allocation technique of the algorithm.

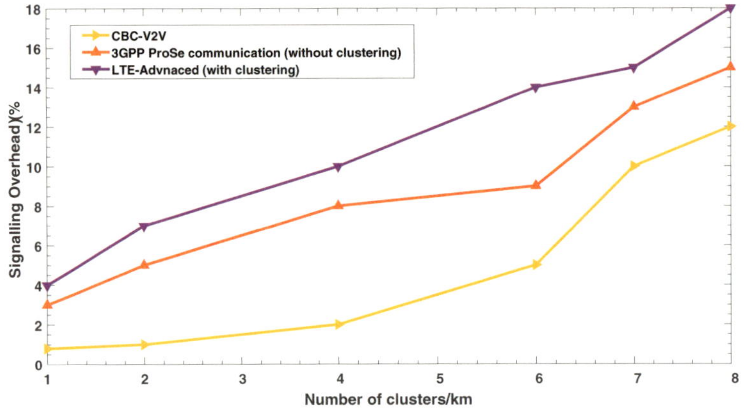

Figure 14.
Performance comparison in terms of signaling overhead.

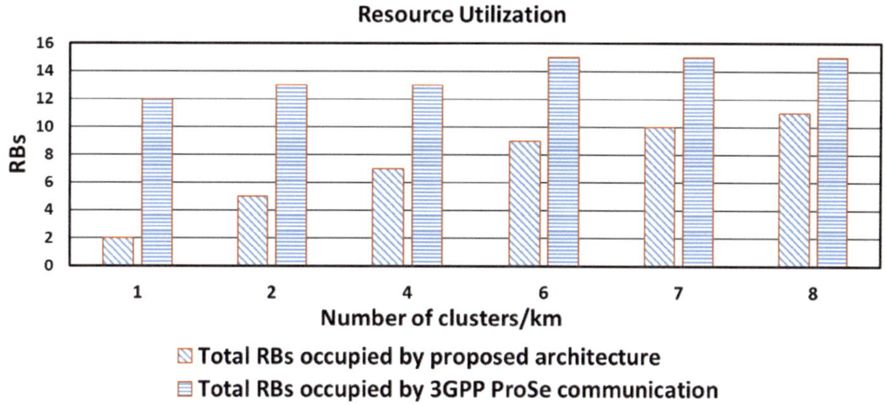

Figure 15.
Performance comparison in terms of total occupied RBs.

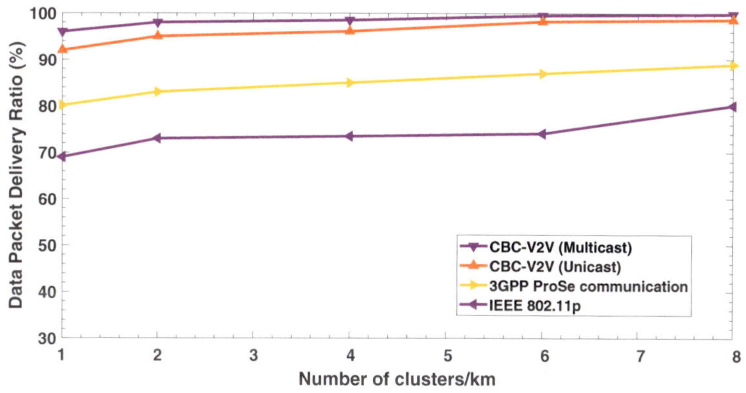

Figure 16.
Performance comparison in terms of data packet delivery ratio.

Figure 16 shows DPDR of the CBC-V2V and compares it with two existing standard procedures. The DPDR is characterized as the proportion of the total number of received safety packets to the total number of scheduled safety packets. Due to the closer vicinity of vehicles, the DPDR value increases with the number of clusters. The system based on IEEE 802.11p shows the lowest DPDR value because packets are lost due to collisions then the proposed D2D packet communication technique which is contention-free. Subsequently, the packet loss probability is low, due to transmission channel condition.

5.1.2 Total end-to-end delay

The total end-to-end delay (δ_{E2E}) for a transmission of a safety message consists of two major delay components as

$$\delta_{E2E} = \delta_{PD} + \delta_{D2D} \tag{3}$$

where δ_{PD} represents the total delay in peer discovery, which is the time difference between sending a request for registration and receiving an eNodeB response and δ_{D2D} represents the total delay in D2D packet communication, which is the sum of the intra- and inter-cluster delays in communication. To ensure timely delivery of active safety messages, the total end-to-end delay (i.e., δ_{E2E}) of the safety message should be less than the required delivery delay (i.e., *Tsafety*).

Figures 17 and **18** present the delay analysis of the CBC-V2V algorithm as a function of the total number of clusters formed based on the number of vehicles/ km. **Figure 17** evaluates and compares the peer discovery delay of the ESPD with the 3GPP ProSe peer discovery model described in Section 4. In the proposed ESPD, the peer discovery delay is the time taken by each vehicle for successful registration. In the registration response, each vehicle receives its current traffic profile which contains the list directly reachable vehicle in its vicinity. Due to the less resource utilization and minimal control signaling overhead requirement, ESPD shows the lower delay values for the peer discovery task compared to the existing 3GPP ProSe peer discovery model. **Figure 18** shows the overall end-to-end packet delay of the CBC-V2V. The results show that the CBC-V2V outperforms the traditional approaches such as IEEE 802.11p, LTE-D2D, and LTE for the safety message transmission in a VANET.

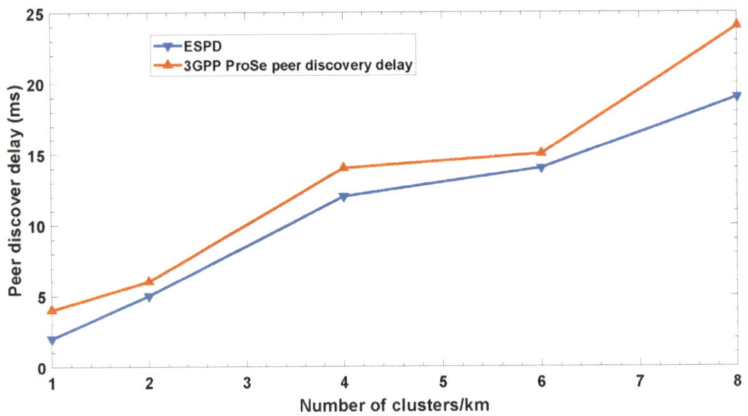

Figure 17.
Performance comparison in terms of peer discovery delay.

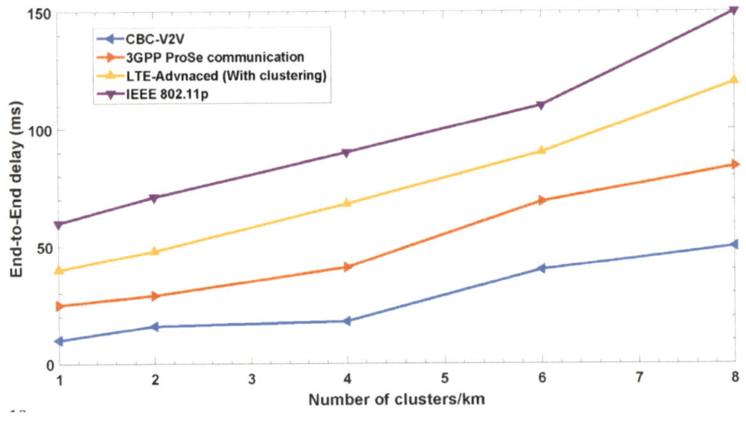

Figure 18.
Performance comparison in terms of E2E packet delay.

6. Conclusion

This chapter has introduced an advanced new cluster-based V2V packet communication architecture combined with an EPC level peer discovery model suitable for vehicular safety applications. The ESPD model reduces the control signaling overhead and end-to-end delay with the awareness of proximity utilizing the GPS information. The CBC-V2V also combines a cluster-based round-robin scheduling technique to distribute the radio resource among the cluster nodes. The CBC-V2V can improve resource utilization and reduce the end-to-end delay to meet the QoS requirements of the safety services in VANETs. Simulation results show that the CBC-V2V offers higher QoS than do the IEEE 802.11p and other LTE networking architectures. The research will be further extended to examine the vehicular network performance in different road terrains and transmission conditions.

Abbreviations

VANET	vehicular ad hoc network
ITS	intelligent transportation system
V2V	vehicle-to-vehicle
V2N	vehicle-to-network
V2P	vehicle-to-pedestrian
V2I	vehicle-to-infrastructure
V2X	vehicle-to-everything
RV	remote vehicle
HV	host vehicle
RSU	road side unit
DENM	decentralized environmental notification messages
STCH	SL traffic channel
SBBCH	SL Broadcast Control Channel
SCI	SL control information
PSSH	physical SL shared channel
PSDCH	physical sidelink discovery channel
ProSe	proximity service
DPDR	data packet delivery ratio
CSMA/CA	carrier-sense multiple access with collision avoidance
SPS	semi-persistent scheduling
LTE	long term evolution
eNB	eNodeB
UE	user equipment
E-UTRAN	Enhanced UMTS Terrestrial Radio Access Network
EPC	evolved packet core
CAM	cooperative awareness message
CBC-V2V	cluster-based cellular vehicle-to-vehicle
ESPD	evolved packet core level sidelink peer discovery
CACC	cooperative adaptive cruise control
SC-FDMA	single-carrier frequency division multiple access
GNSS	global navigation satellite system
EPUID	EPC ProSe subscriber ID
ALUID	application layer user ID
PSS	primary synchronization signal

SSS	secondary synchronization signal
MIB	master information block
SIB2	system information block 2
MRD	maximum reuse distance
FCD	floating car data
CMs	cluster members
CH	cluster head
ICI	inter-carrier interference
TTI	transmission time interval

Author details

Shashank Kumar Gupta*, Jamil Yusuf Khan and Duy Trong Ngo
School of Electrical Engineering and Computing, The University of Newcastle, Callaghan, NSW, Australia

*Address all correspondence to: c3265964@uon.edu.au

References

[1] Chen S, Hu J, Shi Y, Peng Y, Fang J, Zhao R, et al. Vehicle-to-everything (v2x) services supported by LTE-based systems and 5G. IEEE Communications Standards Magazine. 2017;**1**:70-76

[2] Campolo C, Molinaro A, Scopigno R. From todays VANETs to tomorrow's planning and the bets for the day after. Vehicular Communications. 2015;**2**(3):158-171

[3] IEEE Standard for Information Technology. Telecommunications and information exchange between systems local and metropolitan area networks-specific requirements. Part II: Wireless LAN medium access control (MAC) and physical layer (PHY) specifications redline. IEEE Std. 802.11-2012 (Revision of IEEE Std 802.11-2007) Redline. 2012. p. 15229

[4] Teixeira FA, e Silva VF, Leoni JL, Macedo DF, Nogueira JMS. Vehicular networks using the IEEE 802.11p standard: An experimental analysis. Vehicular Communications. 2014;**1**(2): 91-96

[5] Araniti G, Campolo C, Condoluci M, Iera A, Molinaro A. LTE for vehicular networking: A survey. IEEE Communications Magazine. 2013;**51**(5): 148157

[6] Seo H, Lee K, Yasukawa S, Peng Y, Sartori P. LTE evolution for vehicle-to-everything services. IEEE Communications Magazine. June 2016; **54**(6):22-28. DOI: 10.1109/MCOM.2016.7497762

[7] 3GPP TR 22.885. Technical Specification Group Services and System Aspects; Study on LTE Support for V2X Services, Rel. 14, v1.0.0. 2015

[8] 3GPP. Evolved universal terrestrial radio access E-UTRA and evolved universal terrestrial radio access network E-UTRAN; overall description, vol. Stage 2 v14.3.0, (Release 14), no. 3GPP; Technical Report 36.213; 2017

[9] 3GPP TR 36.843. Study on LTE device to device proximity services: Radio aspects Release 12, vol. v12.0.1, Release 12, no. 3GPP; Technical Report 36.843; 2014

[10] 3GPP TS 23.303. 3rd generation partnership project; technical specification group services and system aspects; proximity-based services (ProSe); Stage 2 (Release 15) V15.1.0 (2018-06)

[11] Sun SH, Hu JL, Peng Y, Pan XM, Zhao L, Fang JY. Support for vehicle-to-everything services based on LTE. IEEE Wireless Communications. 2016; **23**(3):48

[12] 3GPP TR 21.914. Technical Specification Group service and system aspects; release 14 description, V14.0.0, 2018-05

[13] 3GPP TS 23.285. Architecture enhancements for V2X services, v16.2.0; 2019

[14] 3GPP TS 36.213. Technical specification group radio access network; Evolved Universal Terrestrial Radio Access (E-UTRA); physical layer procedures, V14.7.0; 2018

[15] 3GPP TS 36.321. Technical specification group radio access network; Evolved Universal Terrestrial Radio Access (E-UTRA); Medium Access Control (MAC) protocol specification, V14.7.0; 2018

[16] 3GPP TS 36.171. Technical specification group radio access network; evolved universal terrestrial radio access (E-UTRA); requirements for support of assisted global navigation

satellite system (A-GNSS), V14.0.0; 2017

[17] 5GAA. V2X Functional and Performance Test Report; Test Procedures and Results. 2018. Available from: http://5gaa.org/wp-content/uploads/2018/11/P-180106-V2X-Functional-and-Performance-TestReport_Final_051118.pdf

[18] Project Cooperative Cars, COCAR [Online]. Available from: http://www.aktivonline.org/english/aktiv-cocar.html

[19] LTE-Connected Cars. NG Connect Program [Online]. Available from: http://ngconnect.org/service-concepts/lte-connected-car/

[20] Sivaraj R, Gopalakrishna AK, Chandra MG, Balamuralidhar P. Qos-enabled group communication in integrated Vanet-LTE heterogeneous wireless networks. In: 2011 IEEE 7th International Conference on Wireless and Mobile Computing, Networking and Communications (WiMob), Conference Proceedings; 2011. p. 1724

[21] Remy G, Senouci S, Jan F, Gourhant Y. LTE4V2X: LTE for a Centralized VANET Organization. In: 2011 IEEE Global Telecommunications Conference—GLOBECOM 2011, Conference Proceedings; 2011. p. 16

[22] Calabuig D, Martn-Sacristn D, Monserrat JF, Botsov M, Gozlvez D. Distribution of road hazard warning messages to distant vehicles in intelligent transport systems. IEEE Transactions on Intelligent Transportation Systems. 2018;**19**(4): 11521165

[23] Toukabri T, Said AM, Abd-Elrahman E, Afifi H. Distributed D2D architecture for ITS services in advanced 4G networks. In: 2015 IEEE 82nd Vehicular Technology Conference (VTC2015-Fall), Conference Proceedings; 2015. p. 17

[24] Molina-Masegosa R, Gozalvez J. LTE-V for sidelink 5G V2X vehicular communications: A new 5G technology for short-range vehicle-to-everything communications. IEEE Vehicular Technology Magazine. 2017;**12**:3039

[25] Min W, Winbjork M, Zhang Z, Blasco R, Do H, Sorrentino S, Belleschi M, Zang Y. Comparison of LTE and DSRC-based connectivity for intelligent transportation systems. In: Proceedings of the IEEE 85th Vehicular Technology Conference (VTC Spring); Sydney, Australia; 2017. p. 47

[26] Nguyen TV, Shailesh P, Sudhir B, Kapil G, Jiang L, Wu Z, Malladi D, Li J. A comparison of cellular vehicle-to-everything and dedicated short range communication. In: Proceedings of the IEEE Vehicular Networking Conference (VNC); Torino, Italy; 27–29 November 2017

[27] Bazzi A, Masini BM, Zanella A, Thibault I. On the performance of IEEE 802.11p and LTE-V2V for the cooperative awareness of connected vehicles. IEEE Transactions on Vehicular Technology. 2017;**66**: 1041910432

[28] 5GAA. An assessment of LTE-V2X (PC5) and 802.11p direct communications technologies for improved road safety in the EU. 2017

[29] 3GPP TS 36.843. Study on LTE device to device proximity services: Radio aspects (Release 12), vol. v12.0.1. 2014

[30] Doumiati S, Artail H. Analytical study of a service discovery system based on an LTE-A D2D implementation. Physical Communication. 2016;**19** (Supplement C):145162

[31] Shinpai Y, Hirkoi H, Satoshi N, Zhao Q. D2D communication in LTE-advanced released 12. NTT DOCOMO Technical Journal. 2015;**17**(2):56-64

[32] Virdis A, Stea G, Nardini G. Simulating LTE/LTE-advanced networks with SimuLTE. In: Obaidat M, Ören T, Kacprzyk J, Filipe J, editors. Simulation and Modeling Methodologies, Technologies and Applications. Advances in Intelligent Systems and Computing, Vol. 402. Cham: Springer; 2015

[33] Sommer C, German R, Dressler F. Bidirectionally coupled network and road traffic simulation for improved IVC analysis. IEEE Transactions on Mobile Computing. 2011;**10**(1):315

[34] Nardini G, Virdis A, Stea G. Simulating device-to-device communications in OMNeT++ with SimuLTE: Scenarios and configurations. In: OMNeT++ Community Summit 2016; Brno, CZ; September 15–16, 2016

[35] Implementations of a Selection of Clustering Algorithms for VANETs, Written in C++ for OMNeT++ [Online]. 2014. Available from: https://github.com/cscooper/ClusterLib

Chapter 7

Healthcare Application-Oriented Non-Lambertian Optical Wireless Communications for B5G&6G

Jupeng Ding, I. Chih-Lin and Jiong Zheng

Abstract

With the continuous improvement of user communication requirements and the rapid development of information services, optical wireless communication (OWC), which has unlimited bandwidth and precise positioning, is widely used in indoor scenes such as healthcare. For healthcare monitoring application, the optical wireless (OW) link using non-Lambertian emission pattern is investigated in the typical mobility scenario. Numerical results show that the potential gain could been provided by the concerned emission pattern to the OW performance uniformity.

Keywords: optical wireless communications, non-Lambertian beams, B5G&6G

1. Introduction

With population aging is emphasized around the world, more attention is paid to the development of the healthcare application with new paradigm. Nevertheless, most of the current health application is based on conventional radio frequency (RF) techniques, such as WiFi (Wireless Fidelity) or UWB (Ultra-Wide Band), and the annoying interference issue frequently degrades the user experience. On the other side, the emerging solid source based optical wireless (OW) technology is consistently investigated to complement the wireless capacity for various healthcare application in EM (electromagnetic) sensitive scenarios [1–5]. Specifically, the validity is examined to achieve the diffuse OW communication between the on-body nodes [6–10].

Up to now, the works of OW healthcare system are still limited to the well-known Lambertian emission pattern which is quite consistent with the conventional solid state sources e.g. LED (Light Emitting Diodes) [11–15]. Nevertheless, there are a number of variations following non-Lambertian emission pattern is still waiting for discussion. In this paper, the typical non-Lambertian OW links is explored in typical healthcare scenario, as shown in **Figure 1** for the first time. And the healthcare OW channel gains comparison are made between the Lambertian & the non-Lambertian configuration.

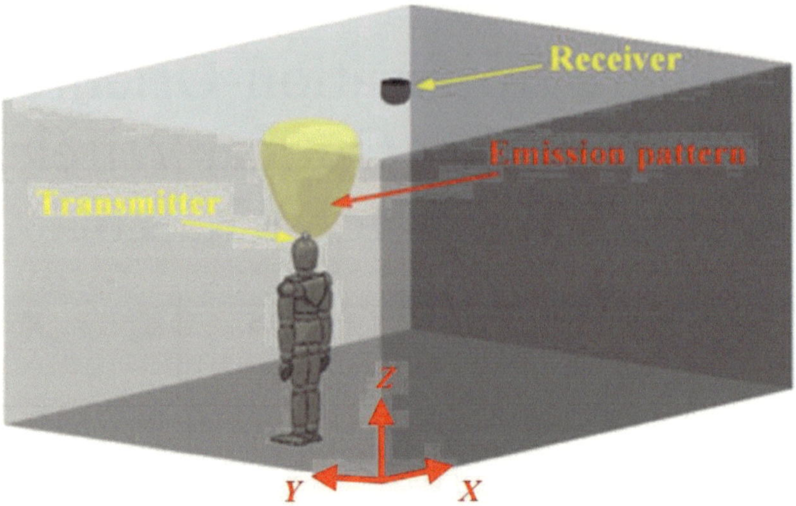

Figure 1.
Typical indoor mobile healthcare scenario.

2. Indoor optical wireless application for healthcare scenario

In this part, the typical non-Lambertian OW links is explored in typical healthcare scenario, as shown in **Figure 1** for the first time. And the healthcare OW channel gains comparison are made between the Lambertian & the non-Lambertian configuration.

2.1 Lambertian & non-Lambertian emission pattern

To the best of our knowledge, in the indoor medical related system shown in the radiation intensity of the transmitter is modeled by the generalized Lambertian pattern as [1, 2]:

$$I_L(\theta) = \frac{m_L + 1}{2\pi} \cos^{m_L}(\theta), \tag{1}$$

where m_L is the Lambertian index and θ is the elevation angle, as shown in **Figure 2a**. At the same time, due to the distinct manufacture process of the solid sources, there are many optical sources could not be characterized by the mentioned Lambertian emission pattern. Typically, one non-Lambertian pattern of the commercially available product i.e. LUXEON® Rebel from Lumileds Philips is presented in **Figure 2b** for comparison.

Following the work of [3, 4], the radiant intensity of this non-Lambertian type could be expressed as:

$$I_{NL}(\theta) = \sum_{i=1}^{2} g_{1i} \exp\left[-\ln 2\left(\frac{|\theta| - g_{2i}}{g_{3i}}\right)^2\right] \tag{2}$$

where $g_{11} = 0.76$, $g_{21} = 0°$, $g_{31} = 29°$, $g_{12} = 1.10$, $g_{22} = 45°$, $g_{32} = 21°$. Obviously, like the Lambertian case, the intensity is independent of the azimuthal angle Φ which basically dominates its symmetry in the far field.

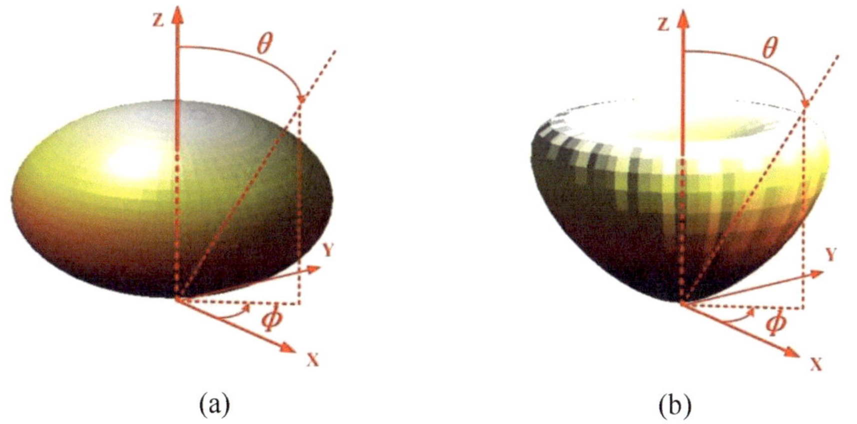

Figure 2.
3D spatial emission patterns in (a) Lambertian type and (b) typical non-Lambertian type.

2.2 Optical wireless link characteristics comparison

In typical indoor healthcare scenario, the OW channel gain form the optical transmitter on the patient to the optical receiver on the ceiling center could be expressed as [1, 2]:

$$H_L = \begin{cases} \dfrac{(m_L+1)A_R}{2\pi d_0^2}\cos^{m_L}(\theta)\cos\psi & \psi < FOV \\ 0 & \psi \geq FOV \end{cases} \tag{3}$$

where A_R is the effective receiver area, d_O is the direct distance from source to optical receiver, and ψ is the angle of incidence on the receiver location. FOV is the field of view of the optical receiver. At the same time, the OW channel gain of the described non-Lambertian emission pattern could be derived as:

$$H_{NL} = \begin{cases} \dfrac{A_R}{d_0^2}\sum_{i=1}^{2} g_{1i}\exp\left[-\ln 2\left(\dfrac{|\theta|-g_{2i}}{g_{3i}}\right)^2\right]\cos\psi & \psi < FOV \\ 0 & \psi \geq FOV \end{cases} \tag{1}$$

For simplifying analysis, the orientation of the optical transmitter is set upward vertically. And the orientation of the optical receiver is set downward vertically. In such situation, emission angle of line of sight (LOS) optical signal equals to the incidence angle at the receiver, i.e. $\theta = \psi$. Such that the optical channel gain of the Lambertian case could be rewritten as:

$$H_L = \begin{cases} \dfrac{(m_L+1)A_R}{2\pi d_0^2}\cos^{m_L+1}(\theta) & \theta < FOV \\ 0 & \theta \geq FOV \end{cases} \tag{5}$$

On the other side, the expression of the non-Lambertian pattern channel gain could been simplified as well:

$$H_{NL} = \begin{cases} \dfrac{A_R}{d_0^2}\sum_{i=1}^{2} g_{1i}\exp\left[-\ln 2\left(\dfrac{|\theta|-g_{2i}}{g_{3i}}\right)^2\right]\cos\theta & \theta < FOV \\ 0 & \theta \geq FOV \end{cases} \tag{6}$$

For fair comparison, the whole emitted optical power of the both emission patterns are normalized to 1 W. The main parameters for the following simulation are included in the **Table 1**. In this Lambertian pattern case, the mobile patient experiences up to 5.77 dB channel gain variation, specifically ranging from −58.71 to −52.94 dB, as shown in **Figure 3a**. Thanks to the intrinsic spatial emission characteristics of the concerned non-Lambertian pattern, the channel gain ranges from −57.79 to −55.26 dB with variation reduced to 2.53 dB. Accordingly, the performance uniformity brought by the pattern replacement could been observed by the probability distribution function (PDF) in **Figure 3b** as well.

Parameters	Value
Length of room	4 [m]
Width of room	3 [m]
Height of room	2.5 [m]
Height of optical receiver	2.5 [m]
Height of optical transmitter	1.8 [m]
Detection area of receiver	1 cm^2
Field of view	85°

Table 1.
Parameters for simulation.

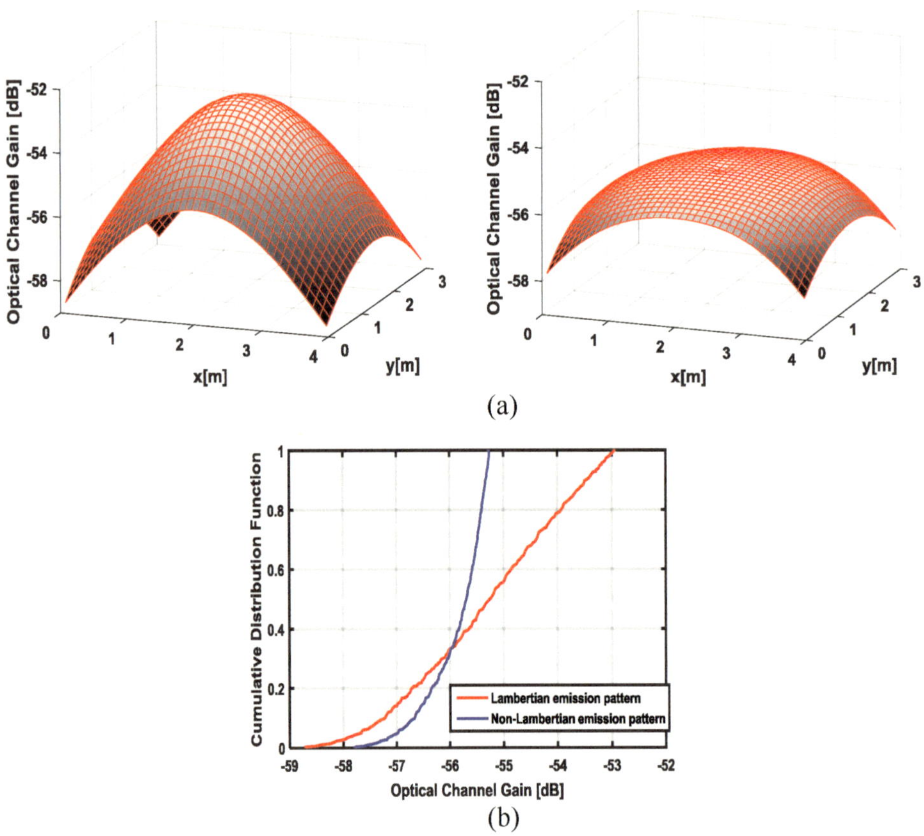

Figure 3.
Optical channel gain comparison in (a) spatial distribution and (b) PDF statistics.

3. Conclusion

The high bandwidth, abundant spectrum resources and high confidentiality of wireless optical communication are suitable for 5G and B5G communication systems. With the rapid development of OWC technology, discussions on different beam characteristics and active research will be unprecedentedly released. In this study, the potential channel gain induced by the non-Lambertian beam is investigated in typical healthcare scenario. The results show that the channel gain fluctuation could be reduced up to about 3.24 dB, with constant transmitted optical power.

Acknowledgements

This work is supported by National Natural Science Foundation of China (Grants No. 62061043), Natural Science Foundation of the Xinjiang Uygur Autonomous Region (Grants No. 2019D01C020), High-level Talents Introduction Project in Xinjiang Uygur Autonomous Region (Grants No. 042419004), Tianchi Doctor Program of the Xinjiang Uygur Autonomous Region (Grants No. 04231200728), Natural Science Foundation of Xinjiang University (Grants No. 62031224624), National Natural Science Foundation of China (Grants No. 61401420), and Tianshan Cedar Project of Xinjiang Uygur Autonomous Region (Grants No. 202101528).

Conflict of interest

The authors declare no conflict of interest.

Author details

Jupeng Ding[1]*, I. Chih-Lin[2] and Jiong Zheng[1]

1 Key Laboratory of Signal Detection and Processing in Xinjiang Uygur Autonomous Region, School of Information Science and Engineering, Xinjiang University, Urumqi, Xinjiang, China

2 China Mobile Research Institute, Beijing, China

*Address all correspondence to: jupeng7778@163.com

References

[1] P. Toumieux, L. Chevalier, S. Sahuguède and A. Julien-Vergonjanne, "Optical wireless connected objects for healthcare," Healthcare Technology Letters, vol. 2, no. 5, pp. 118-122, 10 2015.

[2] Torkestani S S, Sahuguede S, Julien-Vergonjanne A, Cances J P. Indoor optical wireless system dedicated to healthcare application in a hospital. IET Communications, 2012; 6(5):541-547. DOI: 10.1049/iet-com.2010.1116.

[3] Moreno I, Sun C C. Modeling the radiation pattern of LEDs. Optics Express, 2008; 16(3): 1808-1819. DOI: 10.1364/OE.16.001808.

[4] Ding J, I C, Xu Z. Indoor optical wireless channel characteristics with distinct source radiation patterns. IEEE Photonics Journal, 2016; 8(1): 1-15. DOI: 10.1109/JPHOT.2015.250842

[5] Pham T, Atsushi K, Keizo I, Toshimasa U, Naokatsu. Hybrid optical wireless-mmWave: ultra-high-speed indoor communications for beyond 5G. Processing of the IEEE INFOCOM 2019- IEEE Conference on Computer Communications Workshops (INFOCOM WKSHPS), 2019, pp. 1003-1004.

[6] Huang C, Hu S, Alexandropoulos G C, Zappone A, Yuen C, Zhang R, Renzo M D, Debbah M. Holographic MIMO surfaces for 6G wireless networks: opportunities, challenges, and trends. IEEE Wireless Communications, 2020; 27:118-125. DOI: 10.1109/MWC.001.1900534.

[7] Khan M, Chakareski J. Neighbor discovery in a free-space-optical UAV network. Processing of the 2019 IEEE Global Communications Conference (GLOBECOM), Waikoloa, HI, USA, 2019, pp. 1-6.

[8] Nakamura K, Nakagawa S, Matsubara H, Tatsui D, Seki K, Haruyama S, Teraoka F. Development of broadband telecommunications system for railways using laser technology. Electrical Engineering, 2015, 132(5): 666-674. DOI: 10.1541/ieejeiss.132.666.

[9] Mori K, Terada M, Yamaguchi D, Nakamura K, Kaneko K, Teraoka F, Haruyama S. Fast handover mechanism for high data rate ground-to-train free-space optical communication transceiver for internet streaming applications. IEICE Transactions on Communications, 2016; E99.B: 1206-1215. DOI: 10.1587/transcom.2015EBP3326.

[10] Cossu G, Sturniolo A, Messa A, Scaradozzi D, Ciaramella E. Full-fledged 10Base-T ethernet underwater optical wireless communication system. IEEE Journal on Selected Areas in Communications, 2018; 36: 192-202. DOI: 10.1109/JSAC.2017.2774702.

[11] Li Y, Yin H, Ji X, Wu B. Design and implementation of underwater wireless optical communication system with high-speed and full-duplex using blue/green light. Proceedings of the 10th International Conference on Communication Software and Networks, 2018, pp. 99-103.

[12] Du J, Hong X, Wang Y, Xu Z, Zhao W, Lv N, Fei C, He S. A comprehensive performance comparison of DFT-S DMT and QAM-DMT in UOWC system in different water environments. IEEE Photonics Journal, 2019; 13(1):7900211, DOI: 10.1109/JPHOT.2020.3044905.

[13] Mozaffari M, Saad W, Bennis M, Debbah M. Optimal transport theory for power-efficient deployment of unmanned aerial vehicles. Processing of the 2016 IEEE International Conference on Communications (ICC). IEEE, 2016.

[14] Zhang Y, Yang Y, Hu B, Yu L, Hu Z. Average BER and outage probability of the ground-to-train OWC link in turbulence with rain. Optics Communications, 2017, 68: 85-90. DOI: 10.1016/j.optcom.2017.04.034.

[15] Chowdhury M Z, Hasan M K, Shahjalal M, Hossan M T, Jang Y M. Optical wireless hybrid networks: trends, opportunities, challenges, and research directions. IEEE Communications Surveys & Tutorials, 2020; 22:930-966. DOI: 10.1109/COMST.2020.2966855.